Multifaceted Roles of Crystallography in Modern Drug Discovery

NATO Science for Peace and Security Series

This Series presents the results of scientific meetings supported under the NATO Programme: Science for Peace and Security (SPS).

The NATO SPS Programme supports meetings in the following Key Priority areas: (1) Defence Against Terrorism; (2) Countering other Threats to Security and (3) NATO, Partner and Mediterranean Dialogue Country Priorities. The types of meeting supported are generally "Advanced Study Institutes" and "Advanced Research Workshops". The NATO SPS Series collects together the results of these meetings. The meetings are co-organized by scientists from NATO countries and scientists from NATO's "Partner" or "Mediterranean Dialogue" countries. The observations and recommendations made at the meetings, as well as the contents of the volumes in the Series, reflect those of participants and contributors only; they should not necessarily be regarded as reflecting NATO views or policy.

Advanced Study Institutes (ASI) are high-level tutorial courses to convey the latest developments in a subject to an advanced-level audience

Advanced Research Workshops (ARW) are expert meetings where an intense but informal exchange of views at the frontiers of a subject aims at identifying directions for future action

Following a transformation of the programme in 2006 the Series has been re-named and re-organised. Recent volumes on topics not related to security, which result from meetings supported under the programme earlier, may be found in the NATO Science Series.

The Series is published by IOS Press, Amsterdam, and Springer, Dordrecht, in conjunction with the NATO Emerging Security Challenges Division.

Sub-Series

A.	Chemistry and Biology	Springer
B.	Physics and Biophysics	Springer
C.	Environmental Security	Springer
D.	Information and Communication Security	IOS Press
E.	Human and Societal Dynamics	IOS Press

http://www.nato.int/science
http://www.springer.com
http://www.iospress.nl

Series A: Chemistry and Biology

Multifaceted Roles of Crystallography in Modern Drug Discovery

edited by

Giovanna Scapin
Structural Chemistry Department
Merck & Co., Inc.
Kenilworth, NJ, USA

Disha Patel
Center for Advanced Biotechnology and Medicine,
and Rutgers University
Piscataway, NJ, USA

and

Eddy Arnold
Center for Advanced Biotechnology and Medicine,
and Rutgers University
Piscataway, NJ, USA

 Springer

Published in Cooperation with NATO Emerging Security Challenges Division

Proceedings of the NATO Advanced Study Institute on
Crystallography in Drug Discovery: a Tool against CBRN Agents
Erice, Italy
30 May – 8 June, 2014

Library of Congress Control Number: 2015933681

ISBN 978-94-017-9724-5 (PB)
ISBN 978-94-017-9718-4 (HB)
ISBN 978-94-017-9719-1 (e-book)
DOI 10.1007/978-94-017-9719-1

Published by Springer,
P.O. Box 17, 3300 AA Dordrecht, The Netherlands.

www.springer.com

Printed on acid-free paper

Preface

This volume comprises papers presented at the 2014 edition of the "Crystallography of Molecular Biology" series. These courses are part of a wide-ranging set of crystallography courses held since 1974 in the hilltop town of Erice, Italy, at the Ettore Majorana Foundation and Centre for Scientific Culture. This series of courses is renowned for bringing leaders in the fields of macromolecular crystallography and biomedicine together with highly motivated students in a warm and informal atmosphere, which encourages a high level of interactions. Lecturers were chosen from world leaders in the fields of structure-based drug design, biochemistry, biophysics, bioinformatics, computational chemistry, and structural biology, and all made great efforts to present cutting-edge science at a level accessible to participants with limited experience. Most presented two lectures, one focused on methodology and another illustrating the structural insights that can be obtained using their methods.

The course included plenary lectures, as well as talks chosen from poster abstracts submitted by participants. Lectures covered a wide-range of topics including targeting protein-protein interactions, structure determination of G-protein coupled receptors and other membrane proteins, evolution of biopharmaceuticals, understanding of epigenetic processes, and targeting kinases and ribosomes for drug development. Protein-ligand interactions were discussed from multiple perspectives; from the use of high-resolution protein structures to demonstrate the importance of water molecules and protons in protein-ligand interactions to the detailed thermodynamic and kinetic studies that can be performed to fully understand the biochemistry of the interactions. Different approaches to drug discovery and development were highlighted by several talks, including drug discovery and design in pharmaceutical industries and the use of protein engineering and crystallographic fragment screening. Several talks addressed the problem of drug resistance and the steps that can be taken to prevent or minimize it. Discussion was not solely limited to experimental approaches: novel computational techniques for structural bioinformatics, lead generation, prediction of protein aggregation, and the use of molecular modelling and data mining to improve drug discovery success rate were also described. The growing importance and evolving roles of structural databases, essential to ensure all data are properly archived and accessible to the public, were also discussed.

The real organizational work of the course was done by Paola Spadon and Annalisa Guerri who, between them, found most of the funding, corresponded with and coordinated the participant selection and together with the team of 'orange scarves' (Fabio Nicoli, Giovanna Avella, Julia Magliozzo, Giancarlo Tria, and Francesca Vallese) created a warm and welcoming environment in which students were free to engage in scientific discussion with their peers and scientific experts. Erin Bolstad played an essential role, organizing all the computing facilities necessary to conduct tutorials and demonstrations, and providing, together with Fred Boyle, Gianni Grassi, and the Ettore Majorana Center staff, superb IT support. The scientific organizers were Eddy Arnold, Richard Pauptit, Giovanna Scapin, and Robert Stroud. Sir Tom Blundell has been the Director of the International School for Crystallography since 1982.

The course was financed by NATO, and we gratefully acknowledge the NATO Science Committee for the continuous support. Generous financial support was also received from the European Crystallographic Association, the International Union of Crystallography, the Organization for the Prohibition of Chemical Weapons, the Cambridge Crystallographic Data Center, Beryllium, Rigaku, Art Robbins Instruments, and Panalytical.

Kenilworth, NJ, USA   Giovanna Scapin
Piscataway, NJ, USA   Disha Patel
Piscataway, NJ, USA   Eddy Arnold

Contents

Contributors

Eddy Arnold Center for Advanced Biotechnology and Medicine, Rutgers University, Piscataway, NJ, USA

David B. Ascher Department of Biochemistry, Sanger Building, University of Cambridge, Cambridge, UK

Joseph D. Bauman Center for Advanced Biotechnology and Medicine, Rutgers University, Piscataway, NJ, USA

Helen M. Berman RCSB Protein Data Bank, Department of Chemistry and Chemical Biology and Center for Integrative Proteomics Research, Rutgers, The State University of New Jersey, Piscataway, NJ, USA

Tom L. Blundell Department of Biochemistry, Sanger Building, University of Cambridge, Cambridge, UK

Gérard Bricogne Global Phasing Ltd., Cambridge, UK

Chun-wa Chung GlaxoSmithKline Research & Development, Stevenage, Hertfordshire, UK

Miles Congreve Heptares Therapeutics Ltd., Biopark, Welwyn Garden City, UK

U. Helena Danielson Department of Chemistry - BMC, Uppsala University, Uppsala, Sweden

Kalyan Das Center for Advanced Biotechnology and Medicine, Rutgers University, Piscataway, NJ, USA

Andrew S. Doré Heptares Therapeutics Ltd., Biopark, Welwyn Garden City, UK

Shuchismita Dutta RCSB Protein Data Bank, Department of Chemistry and Chemical Biology and Center for Integrative Proteomics Research, Rutgers, The State University of New Jersey, Piscataway, NJ, USA

Ricardo Graña-Montes Institut de Biotecnologia i Biomedicina, Universitat Autònoma de Barcelona, Barcelona, Spain

Colin Groom Cambridge Crystallographic Data Centre, Cambridge, UK

Michael M. Hann GlaxoSmithKline Research and Development, Chemical Sciences, Molecular Discovery Research, Stevenage, UK

Alicia Higueruelo Department of Biochemistry, Sanger Building, University of Cambridge, Cambridge, UK

Eduardo Howard Department of Biophysics, IFLYSIB, CONICET- UNLP, La Plata, Argentina

Ali Jazayeri Heptares Therapeutics Ltd., Biopark, Welwyn Garden City, UK

Harry C. Jubb Department of Biochemistry, Sanger Building, University of Cambridge, Cambridge, UK

Gerhard Klebe Department of Pharmaceutical Chemistry, University of Marburg, Marburg, Germany

Andrew C. Kruse Harvard Medical School, BCMP, Boston, MA, USA

Sergio E. Martinez Center for Advanced Biotechnology and Medicine, Rutgers University, Piscataway, NJ, USA

Rebecca Nonoo Heptares Therapeutics Ltd., Biopark, Welwyn Garden City, UK

Takashi Ochi Department of Biochemistry, Sanger Building, University of Cambridge, Cambridge, UK

Disha Patel Center for Advanced Biotechnology and Medicine, Rutgers University, Piscataway, NJ, USA

Richard Pauptit Discovery Sciences, AstraZeneca, Cheshire, UK

Douglas E.V. Pires Department of Biochemistry, Sanger Building, University of Cambridge, Cambridge, UK

Alberto Podjarny Department of Integrative Biology, IGBMC, CNRS-INSERM-UDS, Illkirch, France

Andreas Prlić RCSB Protein Data Bank, San Diego Supercomputer Center, University of California San Diego, La Jolla, CA, USA

Peter W. Rose RCSB Protein Data Bank, San Diego Supercomputer Center, University of California San Diego, La Jolla, CA, USA

Giovanna Scapin Structural Chemistry Department, Merck & Co., Inc., Kenilworth, NJ, USA

Oliver S. Smart Global Phasing Ltd., Cambridge, UK

SmartSci Limited, St John's Innovation Centre, Cambridge, UK

Salvador Ventura Institut de Biotecnologia i Biomedicina, Universitat Autònoma de Barcelona, Barcelona, Spain

Christine Zardecki RCSB Protein Data Bank, Department of Chemistry and Chemical Biology and Center for Integrative Proteomics Research, Rutgers, The State University of New Jersey, Piscataway, NJ, USA

Chapter 1
Engineering G Protein-Coupled Receptors for Drug Design

Miles Congreve, Andrew S. Doré, Ali Jazayeri, and Rebecca Nonoo

Abstract G protein-coupled receptors (GPCRs) play a crucial role in many diseases and are the site of action of 25–30 % of current drugs (Overington et al., Nat Rev Drug Discov 5(12):993–996, 2006). As such GPCRs represent a major area of interest for the pharmaceutical industry. Despite the rich history of this target class there remain many opportunities for clinical intervention and there is a scarcity of high quality drug-like molecules for many receptors. High-throughput screening has often failed to unlock the potential of members of this superfamily and new, complementary approaches to GPCR drug discovery are required. However, the instability of GPCRs when removed from the cell membrane has severely limited the application of the techniques of structure-based and fragment-based drug discovery. The Heptares approach is successfully overcoming this fundamental challenge and facilitates both biophysical and biochemical fragment screening and also the generation of structural information. Heptares uses its StaR® technology to thermostabilise GPCRs using mutations in precisely defined biologically-relevant conformations (Robertson et al., Neuropharmacology 60(1):36–44, 2011). StaR proteins are amenable to techniques that cannot be readily used with wild-type GPCRs, including fragment screening, biophysical kinetic profiling and X-ray crystallography. Crystal structures of multiple GPCRs have been solved using this approach in the last 5 years (Doré et al., Structure 19(9):1283–1293, 2011; Doré et al., Nature 511:557–562, 2014; Hollenstein et al., Nature 499(7459):438–443, 2013).

A description of the StaR engineering approach, with several examples of how it has been applied, will be presented here. Three examples of how the StaR technology has impacted drug design are outlined. Firstly, X-ray structures of the adenosine $A_{2A}R$ StaR in the antagonist conformation have allowed identification of a clinical candidate progressing into phase 1 clinical trials for the treatment of attention deficit hyperactivity disorder (ADHD) and with potential for the treatment

M. Congreve (✉) • A.S. Doré • A. Jazayeri • R. Nonoo
Heptares Therapeutics Ltd., Biopark, Welwyn Garden City, UK
e-mail: miles.congreve@heptares.com

© Springer Science+Business Media Dordrecht 2015
G. Scapin et al. (eds.), *Multifaceted Roles of Crystallography in Modern Drug Discovery*, NATO Science for Peace and Security Series A: Chemistry and Biology, DOI 10.1007/978-94-017-9719-1_1

1

of Parkinson's disease. Secondly, the first Class B GPCR to be crystallized with a small molecule antagonist ligand, Corticotropin-releasing factor 1 (CRF$_1$R) is presented. Finally, a Class C GPCR, metabotropic glutamate receptor 5 (mGlu$_5$) has been crystallised with a negative allosteric modulator (NAM). Again in this case fragment and structure based drug design has been used to identify a pre-clinical candidate for the potential treatment of a range of CNS disorders.

1.1 Introduction

G protein-coupled receptors (GPCRs) are a major class of cell surface proteins responding to hormones and neurotransmitters [1]. Many CNS drugs mediate their activity through GPCRs including olanzapine for schizophrenia, opioids for pain and dopamine agonists for Parkinson's disease. However there have been few recent successes, with only 10 new GPCRs being drugged in the last decade [2]. New GPCRs of interest including the receptors for neuropeptides and lipids have proved more difficult to find modulators with good in vivo activity alongside the required selectivity, safety and pharmacokinetics. High throughput screening is frequently used to discover starting points for GPCR drug discovery, however often the leads derived from this approach have a high molecular weight and lipophilicity [3]. Such compounds often fail during clinical development due to poor pharmacokinetics and off target toxicity.

The use of modern biophysical and structure-based techniques in drug discovery has been very effectively applied to soluble targets such as enzymes [4]. Structural biology of membrane proteins including GPCRs has been hampered by many factors including low levels of expression, heterogeneity within purified protein populations, flexibility, the existence of different conformational states, instability when removed from the membrane and lack of accessible protein surface for contacts in crystal lattice formation. A range of protein engineering strategies have now been developed which address many of these issues and have resulted in the structures of over 15 GPCRs being solved in the last few years [5].

Heptares Therapeutics uses the technique of conformational thermostabilisation to generate GPCRs with an increased stability in a chosen conformational state, which can be used to facilitate crystallisation and structure determination [6]. These stabilised receptors are known as StaR proteins. StaR proteins have been produced for over 30 different GPCRs in different sub families including family A, B, C and F.

Once the structure is solved a model of the receptor structure is obtained and this is used for virtual screening and structure based drug discovery. X-ray structures are highly enabling for GPCR drug discovery as they allow drug candidates to be designed which fit efficiently into the ligand binding pocket. Compounds can be optimised for affinity, kinetics and selectivity using structure based approaches.

1.2   Results and Discussion

1.2.1   Engineering StaR® Constructs

The process of StaR generation begins with establishing a thermal stability assay. In its simplest form this involves using a radiolabelled ligand of defined pharmacology to measure levels of binding to detergent solubilised receptor at different temperatures. The Tm is defined as the temperature at which 50 % ligand binding can be supported. The use of radiolabelled ligand as a read-out of receptor integrity not only provides a sensitive and quantitative platform it also drives the receptor equilibrium to the desired conformation. Following the generation of a mutation library, the stability of every mutation is measured and compared to the wild-type receptor. Mutations that increase the stability of the receptor are then combined to identify synergistic sets that maximally increase the thermal stability. Such increases can be as much as 30 °C with the introduction of fewer than 10 point mutations. The application of this process results in thermally stable receptors that exhibit resistance to denaturation and aggregation in detergent and more importantly primarily occupy a single conformation which is confirmed with full pharmacological characterisations. The combination of thermal stability and unique conformation is termed conformational stabilisation which ultimately results in the generation of a highly pure and homogeneous protein preparation, significantly increasing the chances of crystallisation and structure determination [6].

1.2.2   FSEC

Fluorescence size exclusion chromatography (FSEC) couples the traditional size exclusion chromatography method with GFP spectroscopy, thus allowing assessment of the biochemical properties of GFP-tagged receptors in crude lysate [7]. The shape of the peak eluting from the column is usually indicative of the quality of the protein. A sharp single peak shows a monodisperse homogeneous protein suitable for crystallisation. Native GPCRs generally show a broad profile with several subpeaks. This is improved by the addition of a suitable conformationally selective ligand, inclusion of thermostabilising mutations and modifications to the protein at the N and C termini such as truncations. FSEC is used to monitor the biochemical properties of StaR proteins during optimisation.

1.2.3   Further Protein Engineering

Prior to crystallisation the protein is optimised by making truncations and removing sites for post-translational modification. GPCRs are usually expressed in the baculovirus insect cell system and purified on an IMAC nickel affinity column using a His-tag at the C terminus of the protein. Other expression systems can also be used for structural studies with GPCRs including mammalian cells [8].

1.2.4   Biophysics and Other Screening

Heptares typically carries out fragment screening to identify hits for further chemical elaboration; fragments are very small compounds typically 100–300 Da in size [9, 10]. Use of the conformational thermostabilisation approach has enabled StaR proteins to be studied in a range of biophysical and other investigations of compound binding, with actual or potential application to fragment screening and/or hit confirmation. This includes surface plasmon resonance (SPR), Target-Immobilized NMR Spectroscopy (TINS) and High Concentration Screening (HCS) by radioligand binding [11].

1.2.5   Biophysical Mapping^{TM}

In this SPR approach, a panel of StaR proteins with individual mutations introduced in the ligand binding site are generated and then probed with a selection of compounds. The purpose of this is to generate a body of site directed mutagenesis (SDM) data to help guide efforts to dock the test compounds, thereby understanding their binding modes to enable SBDD. In addition to monitoring changes in affinities, these Biophysical Mapping (BPM) experiments can highlight changes in association and dissociation rate constants [12]. BPM can be very useful to confirm a compound is binding specifically to the target receptor and, if so, which binding site it engages with. BPM has advantages over the usual SDM methodology because it is a direct binding method, i.e. it does not require competition with a probe ligand, such as a radiochemical. This makes interpretation of the results more straightforward as it is possible to assess directly the differences in binding induced by each mutation for each compound studied. It is also a relatively high throughput approach (compared to SDM) because for each mutant StaR construct a library of compounds can be screened, although in practice 10–30 molecules is more typically used. The results from BPM experiments can help facilitate homology modelling of the receptor and therefore enable virtual (in silico) screening as a source of hit molecules for the GPCR target.

1.2.6   SPR Fragment Screening

This SPR approach involves screening of libraries of fragments, which can be tested at relatively high concentrations (up to 500 μM) against the target StaR protein. Currently, a library of 3,000–4,000 compounds can be screened routinely in 2–3 weeks on a Biacore T200 instrument. The SPR experiment is configured with the protein on the chip and compounds injected as discrete samples [11].

1.2.7 TINS Screening

Target-immobilized NMR screening (TINS) is an NMR method developed to detect weak interactions between proteins and small molecules and has typically been used with soluble proteins. Due to the inherent instability of membrane proteins in solution the TINS method cannot be readily applied to GPCRs. However, combining the TINS approach with the enhanced stability of StaR proteins has allowed for fragment screening and hit identification [13].

1.2.8 High Concentration Screening

Displacement of a radioligand from its binding site is a robust and extremely well characterised approach for HTS screening and has been successfully applied to fragment screening of both GPCRs and soluble proteins [14]. In our laboratories we have extensively used this method for GPCR targets for which ligands are known and a radioligand binding assay can therefore be configured. To facilitate the approach, a number of new GPCR radiochemicals have been developed, allowing HCS screening to be carried out using a library of 3,000–4,000 fragments. StaR constructs as reagents for screening have the advantage of allowing higher concentrations of DMSO to be used in the assay buffer, enabling screening of libraries at high concentrations (such as 300 μM) whilst ensuring the potential ligands remain in solution. Virtual screening derived compound libraries are also typically screened by HCS.

1.2.9 Protein Crystallisation Techniques

Crystallisation of GPCRs is highly challenging due to their instability when removed from the cell membrane and their intrinsic flexibility [15]. Crystallisation screens are set up in a wide range of different detergent conditions. Two approaches are routinely used: vapour diffusion in short chain detergents and lipidic cubic phase (LCP) crystallisation. The lipidic cubic phase is a lipid matrix, which produces a more native, membrane-like environment to aid crystallogenesis [16–18]. Extensive screening and optimisation of crystallisation conditions is then required to obtain the best diffracting crystals. Crystallisation in LCP is frequently facilitated by making fusion proteins with the GPCR, which increase the available surface for mediating crystal contacts [1, 19]. For example T4 lysozyme may be fused into the third intracellular loop of the receptor [20–23]. Antibodies may also be included as crystallisation chaperones [24, 25]. The resulting GPCR crystals are relatively small (10–20 μm) and must be taken to specialised microfocus beamlines to be analysed using synchrotron radiation. Selection of the ligand for co-crystallisation studies is also important. In the absence of thermostabilising mutations a high affinity stabilising ligand is absolutely required to obtain crystals [5]. The presence of thermostabilising mutations reduces this requirement and enables multiple co-crystal structures of weaker ligands to be obtained. This is particularly important when using co-structures during the lead optimisation stage of drug discovery.

Example 1: Adenosine $A_{2A}R$ Antagonist X-Ray Crystal Structures and Identification of a Clinical Candidate Adenosine A_{2A} receptor antagonism has been explored for some time as a potential mechanism for the treatment of Parkinson's disease (PD) [26]. Antagonism of the A_{2A} receptor provides an alternative pathway to increase D_2 receptor signalling, and $A_{2A}R$ antagonists have been shown to be effective preclinically in animal models of PD, with several agents progressing into clinical studies [27–29]. There is also a growing body of evidence preclinically for the role of A_{2A} receptor antagonists in attention and their possible utility in attention deficit hyperactivity disorder (ADHD) [30].

The $A_{2A}R$ structure was first elucidated in 2008 as a fusion protein with T4 lysozyme (3EML) [20]. The structure was solved at 2.6 Å resolution in an inactive conformation, with ZM241385, an inverse agonist ligand, bound in the orthosteric pocket. Structures of $A_{2A}R$ have since been solved in both active and inactive conformations, using fusion proteins such as T4 lysozyme [20, 31] thermostabilised receptor (StaR construct) [32, 33] or as an antibody co-complex [34]. A high resolution structure (1.8 Å) was recently reported using apocytochrome b_{562}RIL from *E. coli* as a fusion partner to reveal $A_{2A}R$ in an inactive conformation [35].

A highly thermostable adenosine $A_{2A}R$ StaR construct in the antagonist conformation (StaR2), has been crystallised with a number of different ligands, which vary in size and potency. Included within this ligand set is the inverse-agonist ZM241385 (PDB code 3PWH) [32], for which the binding mode to the $A_{2A}R$ StaR2 demonstrates some close similarities to that observed in the T4 lysozyme structure reported by Jaakola (PDB code 3EML). In both structures, the furan ring is seen to interact with Asn253$^{6.55}$ *via* a hydrogen bond, and the triazolotriazine core sits between Phe168 and Ile274$^{7.39}$. The binding modes between the $A_{2A}R$ StaR2 and T4-lysozyme structures deviate most significantly at the phenolic end of the ligand. In the StaR structure, the hydroxyl group forms a hydrogen bond to the backbone carbonyl oxygen of Ala63$^{2.61}$, with the phenol ring bound in a cleft between helices TM1, 2 and 7 (defined by Glu13$^{1.39}$, Ala63$^{2.61}$, Ile66$^{2.64}$, Ser67$^{2.65}$, Leu267$^{7.32}$, Met270$^{7.35}$, Ile274$^{7.39}$, His278$^{7.43}$, and Tyr271$^{7.36}$) [32]. Formation of this hydrogen bond may account, at least in part, for the lower affinity demonstrated by the non-hydroxylated ZM241385 analogue, in addition to the high selectivity for $A_{2A}R$ vs. A_1R [36]. The binding modes for ZM241385 in the T4-lysozyme and antibody structures are closely aligned, and in both structures the phenolic tail of ZM241385 points to solvent. This suggests that there may be more than one binding mode for this ligand.

In addition to ZM241385, $A_{2A}R$ StaR2 has been crystallised with the xanthine derivatives XAC (PDB code 3REY) and caffeine (PDB code 3RFM) [32]. Both the high affinity ligand XAC and the low affinity fragment-sized caffeine show some similarities in binding orientation to ZM241385. Both ligands are seen to form a hydrogen bond from their core heterocycles to Asn253[6.55] as seen for the furan oxygen in ZM241385. Interestingly however, the Asn253[6.55] side chain is seen to undergo a rotation in the XAC structure relative to the ZM241385 structure, allowing the Asn NH_2 donor to interact with the XAC carbonyl oxygen to form a 2.9 Å hydrogen bond. As observed for the triazolotriazine core in ZM241385, the core xanthine heterocycles in both XAC and caffeine form π-stacking interactions with Phe168 and make hydrophobic contacts with Ile274[7.39]. The two propyl appendages on the XAC heterocyclic core extend out towards the bottom of the binding site, to interact with Ile66[2.64] on one side in addition to Asn181[5.42] and Leu85[3.33] on the other. The water-solubilising polar tail of XAC binds in a similar way to ZM241385, in a groove created by rotation of Tyr9[1.35] and Tyr271[7.36], between TMs 1, 2 and 7. This region of the receptor is highly flexible, and the two tyrosine residues adopt different rotameric states in the XAC and ZM241385 structures. The cleft effectively created in the XAC structure is seen to be deeper and narrower than in the 3PWH structure, and interestingly is identical to that seen in the structure with caffeine, despite that caffeine does not extend into this region. This may indicate that this is a low energy receptor conformation. These interactions are consistent with site directed mutagenesis data generated using Biophysical Mapping (BPM), a technique described earlier [12].

A medicinal chemistry project at Heptares, designed to identify $A_{2A}R$ antagonists, led to the discovery of a range of hit molecules, from which a series of 3-amino-1,2,4-triazines were prioritised for significant optimisation efforts. During the optimisation process, several 3-amino-1,2,4-triazine ligands were crystallised with $A_{2A}R$ StaR2 (e.g. **1**: PDB code 3UZA and **2**: PDB code 3UZC), (Scheme 1.1) [37]. The binding modes for both ligands **1** and **2** were observed to show similarities to the heterocyclic core of ZM241385, with the central aminotriazine ring seen to π-stack with Phe168 in addition to forming hydrophobic contacts with Ile274[7.39] (Figs. 1.1b and 1.2). As seen for the furan oxygen in ZM241385, both aminotriazine ligands are seen to hydrogen bond to Asn253[6.55], although these ligands form two hydrogen bonds to Asn253[6.55] rather than the single one observed for ZM241385. The side chain carbonyl oxygen of Asn253[6.55] forms one interaction with the triazine amino group and the Asn253[6.55] side chain NH_2 interacts with the N4 nitrogen

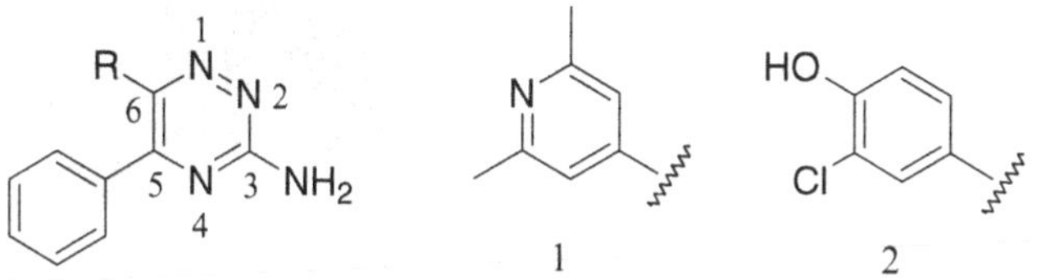

Scheme 1.1 3-Amino-1,2,4-triazine ligands which were crystallised with $A_{2A}R$ StaR2

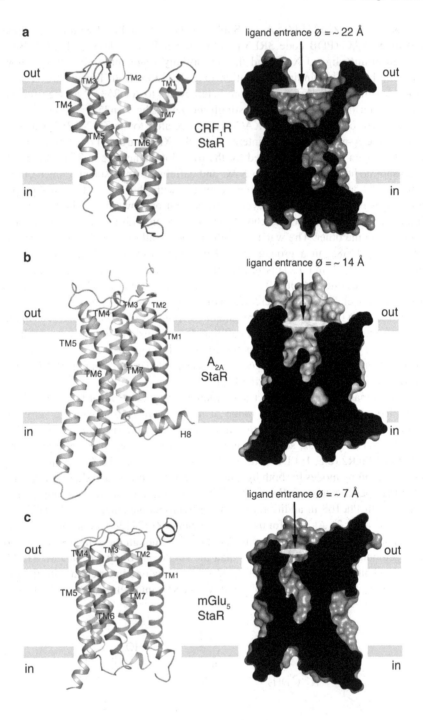

of the triazine ring. A hydrophobic pocket defined by residues $His250^{6.52}$, $Leu85^{3.33}$, $Met177^{5.38}$, $Asn181^{5.42}$ and $Trp246^{6.48}$ accommodates the phenyl ring substituted at the 5-position of the triazine core. In agreement with this, the previously noted BPM study identified that $Leu85^{3.33}$ and $Asn181^{5.42}$ play a significant role in binding of both ligands 1 and 2 [12]. The binding modes for ligands 1 and 2 are most divergent at the position of their differing R-groups. The hydroxyl group present in ligand 2 is able to form a hydrogen bond to $His278^{7.43}$, deep within the receptor ribose pocket, which the pyridyl substituent in 1 is unable to do. Formation of this hydrogen bond to 2 causes the ligand to sit closer to $His278^{7.43}$, resulting in the displacement of the phenol ring atoms relative to the dimethylpyridine ring in 1. However, 1 does occupy the ribose pocket of $A_{2A}R$, as defined by residues $His278^{7.43}$, $Ser277^{7.42}$ Ala (StaR mutation), $Val84^{3.32}$, $Ala63^{2.61}$ and $Ile66^{2.64}$.

Use of this detailed structural information, both for the triazine series itself and also comparisons with other ligand binding modes, allowed SBDD optimisation of this series to potent and sub-type selective molecules in a rapid time frame. Overall, the culmination of efforts in the adenosine $A_{2A}R$ project led to the identification of a clinical candidate molecule, HTL1071, from the 3-amino-1,2,4-triazine chemical series, closely related to ligands 1 and 2 described above. This compound has now completed pre-clinical development and, at the time of writing, is anticipated to enter phase 1 clinical trials shortly.

Example 2: CRF₁R X-Ray Crystal Structure Reveals a Novel Druggable Binding Site The Secretin Class B GPCR subfamily includes 15 receptors for peptide hormone ligands [38], which are important drug targets in a number of human diseases. Examples include the calcitonin gene-related peptide (CGRP) receptor (CLR/RAMP1) for migraine, corticotropin-releasing factor (CRF) receptor for depression and anxiety, glucagon and glucagon-like peptide (GLP) receptors for diabetes and parathyroid peptide hormone (PTH) receptor for osteoporosis [39]. This family of receptors has proven highly challenging to target with small molecules, and in many cases the natural peptide ligand is used as a therapeutic agent [2]. Secretin family receptor proteins characteristically have large N-terminal

---

Fig. 1.1 Cross comparison of StaR structures from Class B, A and C superfamilies. (**a**) *Left* – Ribbon representation of CRF₁R structure (Class B – PDB ID: 4K5Y), viewed parallel to the membrane. Transmembrane (*TM*) helices labeled accordingly, approximate membrane boundaries are shown along with extracellular and intracellular polarity. *Right* – CRF₁R in surface representation slabbed to view the wide opening of the solvent accessible orthosteric/peptide binding pocket (ellipsoid denotes theoretical entrance dimensions at widest point). (**b**) *Left* – Ribbon representation of Adenosine A_{2A} (Class A – PDB ID: 3UZC) structure, viewed parallel to the membrane. *Right* – A_{2A} in surface representation slabbed to view the ligand binding pocket (ellipsoid denotes theoretical entrance dimensions at widest point). (**c**) *Left* – Ribbon representation of mGlu₅ (Class C – PDB ID: 4OO9) structure, viewed parallel to the membrane. *Right* – mGlu₅ in surface representation slabbed to view the narrow entrance to the allosteric pocket (ellipsoid denotes theoretical entrance dimensions at widest point) (Colour figure online)

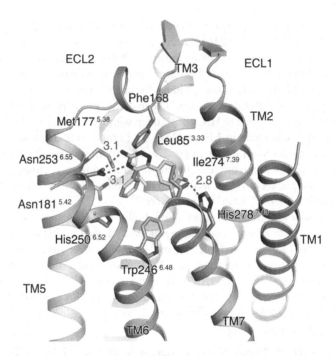

Fig. 1.2 Ligand-binding site of the Adenosine A_{2A} StaR in complex with 4-(3-amino-5-phenyl-1,2,4-triazin-6-yl)-2-chlorophenol. Diagram of ligand interactions in the orthosteric binding pocket viewed from extracellular space. A_{2A} StaR (PDB ID: 3UZC) is shown in ribbon representation; ligand shown as sticks; interacting residues shown as sticks. Hydrogen bonds are depicted as *dashed lines* and their lengths in Å are indicated. Transmembrane helices and extracellular loops labeled accordingly – TM6/TM7 partially removed along with extracellular loop 3 for clarity (Colour figure online)

extracellular domains of 100–160 residues, which are of central importance in receptor activation and ligand binding specificity.

CRF receptors (CRF_1R and CRF_2R) are part of the Secretin family, and bind CRF and urocortin (Ucn 1–3) peptides. Activation of the hypothalamic pituitary axis by the CRF receptors coordinates the body's response to stress [40], influencing physiological functions such as behaviour, appetite control, cardiovascular regulation and immune function. CRF_1 receptor antagonism has been evaluated as a mechanism to treat stress-related disorders such as anxiety and depression, and antagonists are currently under investigation in clinical trials to treat psychiatric indications such as alcoholism [41]. However, clinical trials have not yet shown these agents to be particularly promising, with no evidence to date of a clinically significant benefit. Whether this is caused by the particular design of the clinical trial, failure to sufficiently block the receptor or an invalid mechanism has yet to be determined. The CRF_2 receptor is expressed in both the CNS and peripheral organs. CRF_2R is present in cardiovascular tissue and Ucn 1–3 have shown promise as potent ionotropic agents for the treatment of heart failure [42]. Additionally, in animal models Ucn 1–3 have shown efficacy in the treatment of diabetes and renal failure [43, 44].

Very recent success in the X-ray crystallisation of several GPCRs from the Secretin subfamily has provided invaluable structural information. In 2013, a StaR of the human CRF$_1$ receptor, which preferentially adopts the inactive conformation, was crystallised with small molecule antagonist 2-aryloxy-4-alkylaminopyridine (CP-376395) [45] to obtain the structure of the receptor TMD at 3.0 Å resolution [46]. Concomitantly, the crystal structure of the TMD of the human glucagon receptor (GCGR) in complex with the antagonist 4-[1-(4-cyclohexylphenyl)-3(3-methanesulfonylphenyl)ureidomethyl]-N-(2H-tetrazol-5yl)benzamide (NNC0640) was solved at 3.4 Å resolution [47]. A comparison of the two structures indicates significant conservation of the 7TM helical bundle, especially across TM1-TM5, with the general canonical arrangement of the TM helices similar to that observed in previously determined GPCR structures. However, the top half of the helical bundles in both GCGR and CRF$_1$R adopt conformations, which are more open towards the extracellular space than has been observed for any other GPCR structure previously determined. This results in a 'chalice-like' configuration, where one side is composed of the extracellular halves of TM2-TM5, and the other side by the extracellular halves of TM1, TM6, and TM7. The resultant V-shaped helical bundle features a large solvent-filled orthosteric pocket, which is readily accessible from the extracellular side.

From a drug discovery point of view, the most striking comparison between the CRF$_1$R and GCGR crystal structures can be made regarding the antagonist binding sites. In the CRF$_1$R structure, the antagonist binds more than 15 Å away from the orthosteric site, deep within the cytoplasmic half of the receptor (Figs. 1.1a and 1.3). In contrast, the GCGR structure did not clearly inform the location of a binding site for the antagonist ligand, with the analogous pocket in CRF$_1$R not evident in the GCGR structure. The allosteric small molecule binding site present in the CRF$_1$R structure is displaced more than 7 Å towards the cytoplasm relative to the majority of monoamine class A GPCR orthosteric ligand binding sites, which are found close to or beyond the extracellular boundary of the membrane [48]. The CP-376395 binding pocket in the CRF$_1$ receptor is defined by the residues of TM3, TM5 and TM6, combining hydrophobic and hydrophilic groups which make the pocket suitable to bind small organic drug-like molecules. Class B GPCRs show a remarkably high sequence identity in this region, and of the 14 residues which CP-376395 is seen to directly interact within the CRF$_1$R structure, seven are identical in GCGR. Hydrophobic contacts are made to the antagonist from common residues Phe284$^{5.51\,b}$, Ile290$^{5.57\,b}$, Thr316$^{6.42\,b}$, Leu319$^{6.45\,b}$, Leu323$^{6.49\,b}$ and Gly324$^{6.50\,b}$ whilst Asn283$^{5.50\,b}$ forms an essential receptor-ligand hydrogen bond.

A comparison of the two class B GPCR crystal structures reveals a shift in TM6 toward the membrane in CRF$_1$R relative to GCGR. The shift is particularly evident at the CP-376395 binding site, presumably to account for the presence of the ligand. This suggests that CP-376395 may induce the formation of its own binding pocket, and that in the absence of the ligand the position of TM6 in CRF$_1$R would be more aligned to that of GCGR. CP-376395 binds selectively to CRF$_1$R

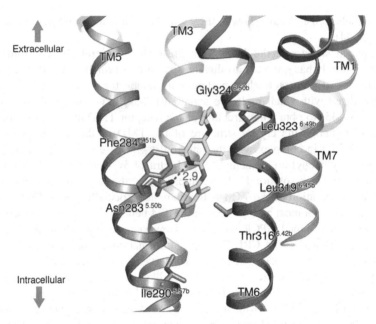

Fig. 1.3 Antagonist-binding site of the CRF$_1$R StaR in complex with CP-376395. Diagram of ligand interactions in the pocket viewed from the membrane. CRF$_1$R StaR (PDB ID: 4K5Y) is shown in ribbon representation; CP-376395 shown as sticks. The single hydrogen bond to the pyridine nitrogen of CP-376395 is depicted as a *dashed line* and its length in Å is indicated. Identical residues between CRF$_1$R and the Glucagon receptor (GCGR) in the antagonist-binding site are shown as sticks. Transmembrane helices and extracellular loops labeled accordingly – the extracellular and intracellular sides of the receptor are indicated by *arrows* – *Left* (Colour figure online)

with no antagonism observed at CRF$_2$R. The origin of this selectivity may arise from the residues in position 3.40b and 5.43b [49]. These residues are histidine and methionine respectively in CRF$_1$R, whilst in CRF$_2$R they are valine and isoleucine. Interestingly, in GCGR these residues are similar to CRF$_2$R (Ile235$^{3.40\,b}$ and Val311$^{5.43\,b}$). Stabilisation of the inactive conformation of the intracellular portion of the CRF$_1$ receptor, potentially preventing the necessary movement of TM6 towards the membrane for docking of the G-protein, may account for the allosteric antagonism observed by CP-376395. Interestingly, the stabilisation of TM6 in such an open conformation by CP-376395 indicates that this small molecule binding pocket could be potentially exploited to design agonist ligands, as suggested by the mutation Thr410$^{6.42\,b}$ Pro, which confers constitutive activity in the PTH1 receptor [50].

As a useful approach to understand receptor druggability, a ligand binding site can be analysed to determine its propensity to bind drug-like molecules [51]. This analysis includes consideration of the binding site shape and size, key physicochemical properties and calculation of the free energies of bound waters. Using this procedure, the binding pockets of H$_1$R, CXCR4, GCGR and CRF$_1$R were compared using GRID (Molecular Discovery) and WaterMap (Schrödinger) [51]. This analysis suggested that the CP-376395 binding region in CRF_1R forms a pocket with promising druggability and has features which are comparable to the orthosteric site in H_1R. This is in comparison to the orthosteric site in CRF_1R, which is open and occupied mainly by bulk solvent, representing a challenging region to target from a drug design perspective. The discovery of a druggable CRF_1R pocket, revealed by X-ray crystallography, therefore opens up a number of exciting possibilities for SBDD.

Example 3: mGlu$_5$ X-Ray Structure with a Negative Allosteric Modulator Ligand Bound Glutamate is the major excitatory neurotransmitter in the brain, exerting its actions through both ionotropic and metabotropic glutamate receptors. There are eight metabotropic glutamate (mGlu) receptors that belong to the class C GPCR family. These receptors can be further divided into three groups according to their sequence similarity, transduction mechanisms and pharmacology. The $mGlu_1$ and $mGlu_5$ receptors belong to group I, are primarily located post-synaptically in the brain and couple through the (G protein) $G_{q/11}$ pathway. Group II mGlu receptors consists of $mGlu_2$ and $mGlu_3$ and group III of $mGlu_4$, $mGlu_6$, $mGlu_7$ and $mGlu_8$. Both group II and III receptors are located pre-synaptically and couple mainly through $G_{i/o}$. Structurally, the mGlu receptors are composed of three regions which consist of the extracellular domain (so-called Venus fly-trap domain), the cysteine rich domain and the transmembrane region. Glutamate is known to bind to a site in the extracellular domain, however, allosteric modulators bind primarily to the transmembrane domain. These modulators can act to enhance (called positive allosteric modulators or PAMs) or decrease (called negative allosteric modulators or NAMs) the activity of glutamate. Overall the mGlu receptors are thought to be involved in the fine tuning of neuronal responses so that inappropriate changes in glutamatergic signalling could potentially play a role in a wide range of human disease processes [52]. There is therefore an opportunity to identify small molecule drugs that serve to specifically modulate the activity of mGlu receptors to treat a variety of neurological and psychiatric disorders.

Very recently the crystal structures of $mGlu_1$ [53] and $mGlu_5$ [54] in complex with NAMs at 2.8 Å and 2.6 Å resolution respectively have been published. The mGlu$_5$ X-ray structure was enabled using the StaR engineering approach outlined earlier. These two structures provide an understanding of TMD configuration for class C GPCRs and in particular describe the atomic details of the allosteric binding site present in both receptors. In both the structures the Venus Fly Trap domain is absent; however, it is known that the TMD region of the mGlu receptors remains a fully functional unit [55]. Global superposition of the TMD domain of $mGlu_5$ and $mGlu_1$ illustrate a high degree of structural conservation of the seven transmembrane helices (RMSD across equivalent $C\alpha = 0.9$ Å). This is important as the structures were solved using different technologies and the degree of similarity strongly suggests that each approach has not introduced any artefacts into the structure solution. This similarity also indicates that in both cases the antagonist conformation of the receptors has been isolated. Comparison of these two structures with class A [56] and B X-ray structures [46] also in the inactive state reveals that Family C receptors adopt a more 'compact' configuration of the helical bundle across the extracellular half of the receptors (Fig. 1.1c). This overall architecture and also the configuration of extracellular loop 2 (between TM4 and 5) produces very restricted entrances to narrow allosteric pockets deep inside the helical bundle. This lack of an obvious entrance to the TMD ligand binding pocket is perhaps consistent with the fact that glutamate does not bind in this region and the pockets are allosteric or 'unnatural'.

In more detail, the binding site of mGlu$_5$ is illustrated in Fig. 1.4, shown in complex with the experimental drug mavoglurant, which spans an upper and lower chamber of the binding pocket. The non-aromatic bicyclic ring system is placed in the upper chamber, surrounded by mainly hydrophobic residues from TM7, 6, 5, and 3. Hydrogen bonds are formed from the ligand to Asn747$^{5.47}$ via the carbamate functional group of mavoglurant and also with the hydroxyl substituent equidistant to the side chains of Ser809$^{7.39}$ and Ser805$^{7.35}$. The ligand contains an alkyne linker which spans a narrow channel into the lower chamber where the 3-methylphenyl ring sits. The lower chamber, depicted in Fig. 1.4, is a complex environment and

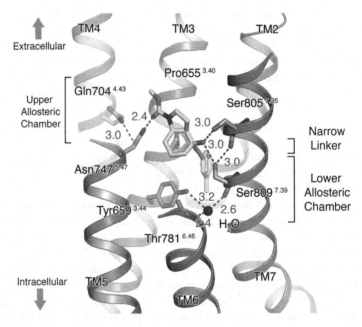

Fig. 1.4 The allosteric modulator binding site of mGlu$_5$. Diagram of ligand interactions in the allosteric pocket viewed from the membrane. The mGlu$_5$ StaR (PDB ID: 4OO9) is shown in ribbon representation; mavoglurant is shown as sticks; interacting residues shown as sticks. Hydrogen bonds between the receptor and ligand are shown as *dashed lines* with distances labeled in Å. The extracellular portions of TM5/TM6 along with extracellular loop 3 have been removed for clarity. Transmembrane helices labeled accordingly – the extracellular and intracellular sides of the receptor are indicated by *arrows* – *Left* (Colour figure online)

may be involved in signalling of this class of receptors. The mGlu receptor crystal structures help to provide a rationalisation of the previously observed narrow SAR and the propensity for 'mode-switching' within chemotypes for mGlu5. Mode switching is the term used when a close analogue of a particular compound changes its pharmacology, e.g. a NAM becoming a PAM. The 3-methyl substituent of mavoglurant is very close to a network of hydrogen bonds involving the side chains of $Tyr659^{3.44}$, $Thr781^{6.46}$, the main-chain of $Ser809^{7.39}$ and an observed water molecule. In similar compounds small changes in this region switches the ligand from a NAM to a neutral binder to a PAM [57]. This may indicate that this buried polar region is part of the activation switch (antagonist state to agonist) for the receptor. Supporting this, it is known that mutation of $Thr781^{6.46}$ and $Ser809^{7.39}$ to alanine changes the pharmacology of alkyne type PAMs [58], again indicating a role of this polar area in the mode switching phenomena [54].

Heptares have used a combination of fragment screening and structure based design to identify a pre-clinical candidate mGlu5 NAM. This compound has the identifier HTL14242 [59]. Although currently unpublished, the binding mode of examples from this chemical series have been solved at high resolution in the $mGlu_5$ StaR X-ray system, described above. Detailed characterisation of this molecule and how it binds to its target receptor will be described at a later date.

In conclusion, the StaR technology developed at Heptares is broadly enabling for SBDD for GPCRs facilitating ligand-receptor co-crystal structure determination for multiple ligands with each receptor being studied. The StaR proteins are well-behaved in a wide array of biophysical techniques allowing fragment based screening approaches and also more detailed characterisation of the binding characteristics of ligands to their receptors. Finally, using the techniques described above, a description of unexpected binding sites for small molecules and how compounds affect the functional signalling of these important drug targets is rapidly being uncovered.

HEPTARES is a registered trademark in the EU, Switzerland, US and Japan; StaR is a registered trademark in the EU and Japan.

References

1. Rosenbaum DM, Rasmussen SGF, Kobilka BK (2009) The structure and function of G-protein-coupled receptors. Nature 459:356–363
2. Congreve M, Langmead CJ, Mason JS et al (2011) Progress in structure based drug design for G protein-coupled receptors. J Med Chem 54(13):4283–4311
3. Leeson PD, Springthorpe B (2007) The influence of drug-like concepts on decision-making in medicinal chemistry. Nat Rev Drug Discov 6:881–890
4. Congreve M, Murray CW, Blundell TL (2005) Structural biology and drug discovery. Drug Discov Today 10(13):895–907
5. Congreve M, Dias JM, Marshall FH (2013) Structure-based drug design for G protein-coupled receptors. In: Progress in medicinal chemistry, vol 53. Elsevier, Amsterdam, p 1
6. Robertson N, Jazayeri A, Errey J et al (2011) The properties of thermostabilised G protein-coupled receptors (StaRs) and their use in drug discovery. Neuropharmacology 60(1):36–44

7. Kawate T, Gouaux E (2006) Fluorescence-detection size-exclusion chromatography for pre-crystallization screening of integral membrane proteins. Stucture 14(4):673–681

8. Lundstrom K, Wagner R, Reinhart C et al (2006) Structural genomics on membrane proteins: comparison of more than 100 GPCRs in 3 expression systems. J Struct Funct Genomics 7(2):77–91

9. Congreve M, Chessari G, Tisi D et al (2008) Recent developments in fragment-based drug discovery. J Med Chem 51(13):3661–3680

10. Erlanson DA, McDowell RS, O'Brien T (2004) Fragment-based drug discovery. J Med Chem 47(14):3463–3482

11. Congreve M, Rich RL, Myszka DG et al (2011) Fragment screening of stabilized G-protein-coupled receptors using biophysical methods. Methods Enzymol 493:115–136

12. Zhukov A, Andrews SP, Errey JC et al (2011) Biophysical mapping of the adenosine A_{2A} receptor. J Med Chem 54(13):4312–4323

13. Chen D, Errey JC, Heitman LH et al (2012) Fragment screening of GPCRs using biophysical methods: identification of ligands of the adenosine A_{2A} receptor with novel biological activity. ACS Chem Biol 7:2064–2073

14. Albert JS, Blomberg N, Breeze AL et al (2007) An integrated approach to fragment-based lead generation: philosophy, strategy and case studies from AstraZeneca's drug discovery programmes. Curr Top Med Chem 7(16):1600–1629

15. Tate CG (2010) Practical considerations of membrane protein instability during purification and crystallisation. Methods Mol Biol 601:187–203

16. Caffrey M, Cherezov V (2009) Crystallizing membrane proteins using lipidic mesophases. Nat Protoc 4:706–731

17. Caffrey M, Li D, Dukkipati A (2012) Membrane protein structure determination using crystallography and lipidic mesophases: recent advances and successes. Biochemistry 51(32):6266–6288

18. Caffrey M (2009) Crystallizing membrane proteins for structure determination: use of lipidic mesophases. Annu Rev Biophys 38:29–51

19. Chun E, Thompson AA, Liu W et al (2012) Fusion partner toolchest for the stabilization and crystallization of G protein-coupled receptors. Structure 20:967–976

20. Jaakola VP, Griffith MT, Hanson MA et al (2008) The 2.6 angstrom crystal structure of a human A_{2A} adenosine receptor bound to an antagonist. Science 322:1211–1217

21. Cherezov V, Rosenbaum DM, Hanson MA et al (2007) High-resolution crystal structure of an engineered human β2-adrenergic G protein-coupled receptor. Science 318:1258–1265

22. Wu B, Chien EY, Mol CD et al (2010) Structures of the CXCR4 chemokine GPCR with small-molecule and cyclic peptide antagonists. Science 330:1066–1071

23. Chien EY, Liu W, Zhao Q et al (2010) Structure of the human dopamine D3 receptor in complex with a D2/D3 selective antagonist. Science 330:1091–1095

24. Rasmussen SG, Choi HJ, Rosenbaum DM et al (2007) Crystal structure of the human β2 adrenergic G-protein-coupled receptor. Nature 450:383–387

25. Day PW, Rasmussen SGF, Parnot C et al (2007) A monoclonal antibody for G protein-coupled receptor crystallography. Nat Methods 4:927–929

26. Gomes CV, Kaster MP, Tomé AR et al (2011) Adenosine receptors and brain diseases: neuroprotection and neurodegeneration. Biochim Biophys Acta 1808:1380–1399

27. Shah U, Hodgson R (2010) Recent progress in the discovery of adenosine A_{2A} receptor antagonists for the treatment of Parkinson's disease. Curr Opin Drug Discov Devel 13:466–480

28. Hodgson RA, Bedard PJ, Varty GB et al (2010) Preladenant, a selective A_{2A} receptor antagonist, is active in primate models of movement disorders. Exp Neurol 225:384–390

29. Salamone JD (2010) Preladenant, a novel adenosine A_{2A} receptor antagonist for the potential treatment of Parkinsonism and other disorders. Drugs 13:723–731

30. Takahashi RN, Pamplona FA, Prediger RD (2008) Adenosine receptor antagonists for cognitive dysfunction: a review of animal studies. Front Biosci 13:2614–2632
31. Xu F, Wu H, Katritch V et al (2011) Structure of an agonist-bound human A_{2A} adenosine receptor. Science 332:322–327
32. Doré AS, Robertson N, Errey JC et al (2011) Structure of the adenosine A_{2A} receptor in complex with ZM241385 and the xanthines XAC and caffeine. Structure 19(9):1283–1293
33. Lebon G, Warne T, Edwards PC et al (2011) Agonist-bound adenosine A_{2A} receptor structures reveal common features of GPCR activation. Nature 474:521–525
34. Hino T, Arakawa T, Iwanari H et al (2012) G-protein-coupled receptor inactivation by an allosteric inverse-agonist antibody. Nature 482:237–240
35. Liu W, Chun E, Thompson AA et al (2012) Structural basis for allosteric regulation of GPCRs by sodium ions. Science 337:232–236
36. de Zwart M, Vollinga RC, Beukers MW et al (1999) Potent antagonists for the human adenosine A_{2B} receptor. Derivative of the triazolotriazine adenosine receptor antagonist ZM241385 with high affinity. Drug Dev Res 48:95–103
37. Congreve M, Andrews SP, Doré AS et al (2012) Discovery of 1,2,4-triazine derivatives as adenosine A_{2A} antagonists using structure based drug design. J Med Chem 55:1898–1903
38. Alexander SPH, Benson HE, Faccenda E et al (2013) The concise guide to pharmacology 2013/14: G protein-coupled receptors. Br J Pharmacol 170:1459–1581
39. Pal K, Melcher K, Xu HE (2012) Structure and mechanism for recognition of peptide hormones by Class B G-protein-coupled receptors. Acta Pharmacol Sin 33:300–311
40. Bale TL, Vale WW (2004) CRF and CRF receptors: role in stress responsivity and other behaviors. Annu Rev Pharmacol Toxicol 44:525–557
41. Zorrilla EP, Heilig M, de Wit H et al (2013) Behavioral, biological, and chemical perspectives on targeting CRF_1 receptor antagonists to treat alcoholism. Drug Alcohol Depend 128:175–186
42. Yang LZ, Tovote P, Rayner M et al (2010) Corticotropin-releasing factor receptors and urocortins, links between the brain and the heart. Eur J Pharmacol 632:1–6
43. Li C, Chen P, Vaughan J et al (2007) Urocortin 3 regulates glucose-stimulated insulin secretion and energy homeostasis. Proc Natl Acad Sci U S A 104:4206–4211
44. Devetzis V, Zarogoulidis P, Kakolyris S et al (2013) The corticotropin releasing factor system in the kidney: perspectives for novel therapeutic intervention in nephrology. Med Res Rev 33:847–872
45. Chen YL, Obach RS, Braselton J et al (2008) 2-Aryloxy-4-alkylaminopyridines: discovery of novel corticotropin-releasing factor 1 antagonists. J Med Chem 51:1385–1392
46. Hollenstein K, Kean J, Bortolato A et al (2013) Structure of class B G-protein-coupled receptor corticotropin-releasing factor receptor 1. Nature 499(7459):438–443
47. Siu FY, He M, de Graaf C et al (2013) Structure of the human glucagon class B G protein-coupled receptor. Nature 499:444–449
48. Venkatakrishnan AJ, Deupi X, Lebon G et al (2013) Molecular signatures of G-protein-coupled receptors. Nature 494:185–194
49. Hoare SRJ, Brown BT, Santos MA et al (2006) Single amino acid residue determinants of non-peptide antagonist binding to the corticotropin-releasing factor$_1$ (CRF_1) receptor. Biochem Pharmacol 72:244–255
50. Schipani E, Jensen GS, Pincus J et al (1997) Constitutive activation of the cyclic adenosine 3',5'-monophosphate signaling pathway by parathyroid hormone (PTH)/PTH-related peptide receptors mutated at the two loci for Jansen's metaphyseal chondrodysplasia. Mol Endocrinol 11:851–858
51. Mason JS, Bortolato A, Congreve M et al (2012) New insights from structural biology into the druggability of G protein-coupled receptors. Trends Pharmacol Sci 33:249–260
52. Yasuhara A, Chaki S (2010) Metabotropic glutamate receptors: potential drug targets for psychiatric disorders. Open Med Chem J 4:20–36

53. Wu H, Wang C, Gregory KJ et al (2014) Structure of a class C GPCR metabotropic glutamate receptor 1 bound to an allosteric modulator. Science 344:58–64
54. Doré AS, Okrasa K, Patel JC et al (2014) Structure of class C GPCR metabotropic glutamate receptor 5 transmembrane domain. Nature 511:557–562
55. Goudet C, Gaven F, Kniazeff J et al (2004) Heptahelical domain of metabotropic glutamate receptor 5 behaves like rhodopsin-like receptors. Proc Natl Acad Sci U S A 101:378–383
56. Palczewski K, Kumasaka T, Hori T et al (2000) Crystal structure of rhodopsin: a G protein-coupled receptor. Science 289:739–745
57. O'Brien JA, Lemaire W, Chen TB et al (2003) A family of highly selective allosteric modulators of the metabotropic glutamate receptor subtype 5. Mol Pharmacol 64:731–740
58. Gregory KJ, Nguyen ED, Reiff SD et al (2013) Probing the metabotropic glutamate receptor 5 (mGlu5) positive allosteric modulator (PAM) binding pocket: discovery of point mutations that engender a "molecular switch" in PAM pharmacology. Mol Pharmacol 83:991–1006
59. Christopher JA, Aves SJ, Bennett KA et al (2014) Fragment-based GPCR drug discovery using a stabilised receptor (StaR): identification of an mGlu5 negative allosteric modulator (NAM) pre-clinical candidate. Abstracts of papers, 247th ACS national meeting & exposition, Dallas, TX, USA, 16–20 March, MEDI-281

Chapter 2
Structural Insights into Activation and Allosteric Modulation of G Protein-Coupled Receptors

Andrew C. Kruse

Abstract G protein-coupled receptors (GPCRs) are cell-surface receptors that regulate neurotransmission, cardiovascular function, metabolic homeostasis, and many other physiological processes. Due to their central role in human physiology, these receptors are among the most important targets of therapeutic drugs, and are they among the most extensively studied integral membrane proteins. To better understand GPCR signaling at a molecular level, we have undertaken structural studies of a prototypical GPCR family, the muscarinic acetylcholine receptors. These studies have led to crystal structures of muscarinic receptors in both inactive and active conformations, as well as the first structure of a GPCR in complex with a drug-like allosteric modulator. In addition, we have recently developed new approaches in combinatorial biology to create protein modulators of GPCR signaling. These studies shed light on the function of muscarinic receptors, and offer insights into the molecular basis for the regulation of GPCR signaling and activation in general.

2.1 Introduction

As critical regulators of human physiology, GPCRs have long been the subject of extensive pharmacological study. More recently, advances in protein expression and purification have made GPCRs amenable to investigation by biochemical methods, spectroscopic studies, and structural biology. GPCR signaling is typically initiated by binding of a ligand, such as a drug, to the extracellular surface of the receptor. This binding event leads to conformational changes in the receptor that allow the GPCR to interact with intracellular signaling proteins including G proteins, arrestins, and GPCR kinases [1]. While many aspects of these complex signaling pathways are understood in great detail, other important questions still remain unanswered, particularly surrounding molecular aspects of ligand binding, selectivity, and receptor activation.

A.C. Kruse (✉)
Harvard Medical School, BCMP, 240 Longwood Ave., Boston, MA 02115, USA
e-mail: Andrew_Kruse@hms.harvard.edu

© Springer Science+Business Media Dordrecht 2015
G. Scapin et al. (eds.), *Multifaceted Roles of Crystallography in Modern Drug Discovery*, NATO Science for Peace and Security Series A: Chemistry and Biology, DOI 10.1007/978-94-017-9719-1_2

To better understand GPCR signaling at a molecular level, we have undertaken structural studies of a prototypical GPCR family, the muscarinic acetylcholine receptors. These receptors were chosen as the subject of this work due to their propensity to bind allosteric modulators, and for their longstanding role as a model system for understanding GPCR signaling in general. Muscarinic receptors are also important targets in the treatment of diseases, including respiratory conditions and neurodegenerative diseases [2]. Muscarinic receptors are also important drug antitargets, since they are responsible for off-target side effects of various drugs, including first-generation antihistamines, which often cause dry mouth due to inhibition of the M_3 muscarinic receptor.

2.2   Results and Discussion

Initially, we sought to understand how small-molecule drugs bind to muscarinic receptors by using X-ray crystallography to determine structures of these receptors bound to different ligands. Unfortunately, muscarinic receptors and other GPCRs are poor candidates for such studies, because they are biochemically unstable and because they possess large, unstructured loops. To address this problem, we followed an approach developed by Rosenbaum et al. [3], and replaced the flexible third intracellular loop of the receptor with T4 lysozyme (T4L), a soluble, crystallizable protein (Fig. 2.1). This approach has been successfully applied to a large number of GPCRs, showing it to be a highly general method. Indeed, a majority of GPCR structures reported to date have been solved using the protein fusion approach [4].

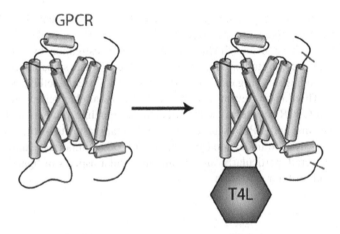

GPCR

T4L

Fig. 2.1 Modification of a GPCR to facilitate crystallization. GPCRs typically contain flexible termini and loops. To address these liabilities, muscarinic receptors had the third intracellular loop replaced with T4 lysozyme (T4L), and had the amino- and carboxy-termini truncated in the case of the M_3 receptor

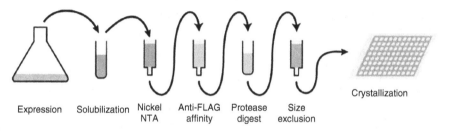

Expression Solubilization Nickel   Anti-FLAG Protease Size   Crystallization
                            NTA   affinity   digest   exclusion

Fig. 2.2 Purification procedure. A typical GPCR purification procedure is diagrammed, showing major steps in the purification. Muscarinic receptors were purified in this manner

We crystallized the T4 lysozyme-fused M_2 and M_3 receptors using the lipidic cubic phase technique [5]. This approach involves reconstitution of the target protein in a lipid bilayer, followed by immersion of the protein/lipid mix in a precipitant solution that diffuses through the sample, promoting crystallization. In the case of the M_2 and M_3 receptors, extensive optimization of the purification procedures was required, leading to the purification procedure outlined in Fig. 2.2. Once homogenous biochemical preparations were obtained, crystallization and structure determination were relatively straightforward, leading to structures of both receptors [6, 7]. As is typical for many lipidic cubic phase crystals, muscarinic receptors formed crystals less than 100 μm in length and required micro-focus data collection methods [8]. The use of micro-focus X-ray sources at the Advanced Photon Source of Argonne National Lab was essential to allowing structure determination from such small crystals, since lipidic cubic phase crystallization drops often contain dozens to hundreds of crystals that cannot be separated from one another. The use of a micro-focus beamline this allows individual crystals to be irradiated sequentially, generating a series of "wedges" of data, each comprising 5°–10° total rotation of the crystal.

Structures of the M_2 and M_3 receptors were solved with the receptors bound to two different antagonists: the clinical drug tiotropium (Spiriva) bound to the M_3 receptor, and the research compound quinuclidinyl benzilate (QNB) bound to the M_2 receptor. The two structures show a high degree of overall similarity, and possess the typical seven transmembrane fold seen in other GPCRs (Fig. 2.3). Closer inspection reveals that M_2 and M_3 receptors possess unique, deeply buried ligand-binding pockets, surrounded by transmembrane helices. The two receptors are virtually identical in terms of ligand binding site conformations (Fig. 2.4), offering an explanation for the difficulties faced by medicinal chemists in designing subtype-selective muscarinic receptor ligands. Key ligand contacts include a charge-charge interaction between $Asp^{3.32}$ (superscripts denote Ballesteros-Weinstein numbering [9]) and the ligand amine, representing a feature conserved in all aminergic GPCRs as well as opioid receptors [10, 11]. The only other polar contact involves $Asn^{6.52}$, which engages in a pair of hydrogen bond interactions with the bound ligand in both muscarinic receptor subtypes. A striking feature of the muscarinic receptor ligand binding pocket is an abundance of aromatic amino acid side chains surrounding the

Fig. 2.3 Overall structure of muscarinic receptors. The overall structures of the muscarinic receptors are shown, with the M_2 and M_3 subtypes superimposed on one another. The overall folds are highly similar both to each other and to other Family A GPCRs

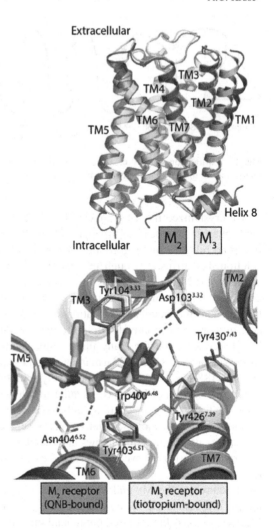

Fig. 2.4 Ligand recognition. The ligand binding pockets of the M_2 and M_3 muscarinic receptors are shown superimposed. Receptor amino acid side chains are shown in *thin sticks*, and the bound ligands (QNB and tiotropium) are shown in *thick sticks*. Residues are numbered according to the M_2 receptor sequence, with Ballesteros-Weinstein numbering in *superscripts*. Polar contacts are indicated with *dotted lines*

positively charge amine. Similar features are seen in the structures of other proteins that bind acetylcholine, suggesting convergent evolution toward a particular ligand recognition mode [6].

The binding site of tiotropium and QNB is referred to as the orthosteric ligand binding site, and this is also the binding pocket for the endogenous agonist acetylcholine. In addition to this site, structures of muscarinic receptors also revealed the existence of a large cavity situated directly above (extracellular to) the orthosteric binding site. This cavity, termed the "extracellular vestibule" is lined by residues that have been previously implicated in binding to allosteric modulators [12]. The location of this cavity is also consistent with the observation that many muscarinic allosteric modulators can slow dissociation of orthosteric ligands, resulting in slowed binding kinetics [12].

To probe the role of the extracellular vestibule in ligand binding, we performed long timescale molecular dynamics simulations of antagonists binding to and dissociating from the muscarinic receptors [7]. These studies suggested that ligands entering or leaving the orthosteric site may pause at the extracellular vestibule, indicating that this second site is also a potential target for small molecule drugs. In fact, experimental data have provided evidence that orthosteric ligands can also act as allosteric modulators at high concentrations [13], presumably binding to the extracellular vestibule and modifying the effects of orthosteric ligands.

More recently, Dror and colleagues have built upon this work and performed a detailed series of simulations and mutagenesis experiments to probe the action of negative allosteric modulators, which bind to the extracellular vestibule [14]. The simulations indicated that negative allosteric modulator binding involves cation-π interactions between the positively charged modulators and aromatic amino acid sidechains surrounding the extracellular vestibule. These simulations offer some of the most detailed views of negative allosteric modulator binding, but to date, no experimentally determined structure of a muscarinic receptor bound to a negative allosteric modulator has been reported.

While the first structures of muscarinic receptors offered important insights into the molecular details of ligand recognition, both structures represent inactive states. We next sought to understand activation of the M_2 muscarinic receptor, as well as regulation of receptor activation by positive allosteric modulators. Initial attempts to crystallize M_2 receptor bound to agonists were unsuccessful, likely due to conformational flexibility in the receptor. Such conformational heterogeneity has been extensively studied in the β_2 adrenergic receptor, a homolog of muscarinic receptors [15]. Based on this hypothesis, we sought to develop stabilizing single domain camelid antibody fragments (called nanobodies) to bind the intracellular side of the receptor and stabilize an active conformation, similar to a previously successful approach developed by Rasmussen et al. [16].

Initial attempts involving a standard immunization and phage display selection and screening approach were unsuccessful, resulting in nanobodies that could bind M_2 receptor but which lacked conformational selectivity. We next turned to a yeast surface display approach, expressing a library of nanobodies on the surface of yeast and staining the cells with M_2 receptor solubilized in detergent. Previously, we had developed this approach to create a high-affinity nanobody specific to the β_2 adrenergic receptor [17]. In the case of the muscarinic receptors, two receptor samples were prepared with distinct fluorophores covalently attached. Each population was bound to either a covalent agonist or a high affinity antagonist. By using a mix of active receptors and inactive receptors labeled with distinct fluorophores (Fig. 2.5), we were able to sort cells by fluorescence activated cell sorting (FACS) to select clones with the desired conformational selectivity [18].

After screening selected clones for the binding and conformational specificity, a high affinity, active-state stabilizing nanobody called Nb9-8 was identified. This nanobody was purified in complex with M_2 receptor and a bound agonist, and crystallized by the lipidic cubic phase method. The resulting structure revealed an active conformation M_2 receptor, showing a rotation of transmembrane helix 6,

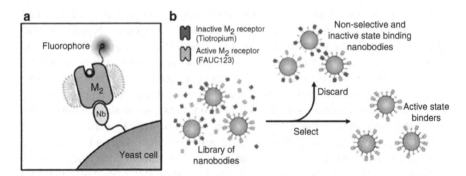

Fig. 2.5 Conformational selection. (**a**) A library of nanobodies was expressed on the surface of Saccharomyces cerevisiae yeast, then stained with purified receptor in detergent. (**b**) This allowed selective isolation of only clones binding to activated receptors in the manner diagrammed here

Fig. 2.6 Activation mechanism. Comparison of inactive- and active-state structures of the M_2 receptor show the overall similar structure with a notable deviation at transmembrane helix 6. The outward rotation of this helix is seen in other activated GPCR structures, suggesting it is a common and conserved feature of GPCR activation

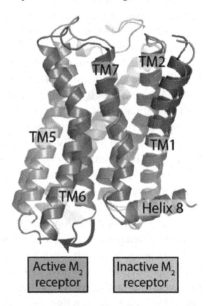

which is a hallmark of GPCR activation (Fig. 2.6). The most surprising feature of this structure is a large rearrangement of the extracellular region of the receptor, which had not previously been observed for other GPCRs. This region of the receptor is known to be the target of allosteric modulators, prompting a second series of crystallization trials with a positive allosteric modulator of agonists, called LY02119620 [19].

Allosteric modulators of GPCR function have become increasingly attractive as potential therapeutics, possessing properties often unachievable with conventional orthosteric ligands [20]. In particular, allosteric modulators can influence signaling while maintaining the native spatiotemporal regulation of agonist release. Allosteric ligands may also show higher selectivity than orthosteric ligands, as they often bind to sites with lower sequence conservation than conventional ligands. This is particularly important in the case of muscarinic acetylcholine receptors, where drugs targeting specific subtypes have not been available. Although drugs targeting the M_1 muscarinic receptor have shown efficacy in the treatment of neurodegenerative diseases, their use in human patients is precluded by side effects due to activation of other muscarinic receptor subtypes. Ligands targeting less conserved allosteric sites are typically more selective for specific receptor subtypes, and may therefore offer a means of developing agonists and antagonists with selectivity toward particular muscarinic receptors.

Despite increased interest in allosteric modulators, no structural information regarding allosteric modulation of GPCRs has been available until recently. Using our engineered nanobody 9-8, we were able to obtain a second crystal structure of activated M_2 receptor, this time bound to the LY02119620 modulator. This structure revealed that the conformation of the receptor is highly similar irrespective of whether or not the modulator is bound, suggesting that the modulator is recognizing a binding site that is essentially "pre-formed" upon receptor activation [18, 21]. By stabilizing this site, the modulator may promote signaling and stabilize agonist binding, thereby accounting for its pharmacological profile (Fig. 2.7).

Taken together, these studies have shown structures of muscarinic receptors in both inactive and active conformations, as well as the first structure of a GPCR in complex with a drug-like allosteric modulator. The use of new approaches in combinatorial biology allowed the identification of nanobody modulators of GPCR signaling, facilitating structural studies of activated muscarinic receptors. Many questions remain, however, and ongoing studies will lead to a more complete understanding of muscarinic receptor function. In particular, the interactions between muscarinic receptors and their effectors (G proteins and arrestins) remain poorly

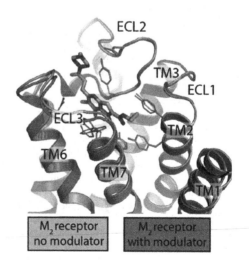

Fig. 2.7 Allosteric modulation. Comparison of active-state M_2 receptor structures with and without bound allosteric modulator LY02119620 (*thick sticks*). In each case, the overall structure of the receptor and the side chain conformations are highly similar (See Ref. [21] for more details)

understood, with little structural information regarding these interactions. In the long term, new insights into the mechanistic basis for muscarinic receptor signaling and GPCR function in general may facilitate the development of new and better therapeutics targeting these important receptors.

References

1. Rosenbaum DM, Rasmussen SG, Kobilka BK (2009) The structure and function of G-protein-coupled receptors. Nature 459(7245):356–363
2. Wess J, Eglen RM, Gautam D (2007) Muscarinic acetylcholine receptors: mutant mice provide new insights for drug development. Nat Rev Drug Discov 6(9):721–733
3. Rasmussen SG et al (2007) Crystal structure of the human beta2 adrenergic G-protein-coupled receptor. Nature 450(7168):383–387
4. Venkatakrishnan AJ et al (2013) Molecular signatures of G-protein-coupled receptors. Nature 494(7436):185–194
5. Caffrey M, Cherezov V (2009) Crystallizing membrane proteins using lipidic mesophases. Nat Protoc 4(5):706–731
6. Haga K et al (2012) Structure of the human M2 muscarinic acetylcholine receptor bound to an antagonist. Nature 482(7386):547–551
7. Kruse AC et al (2012) Structure and dynamics of the M3 muscarinic acetylcholine receptor. Nature 482(7386):552–556
8. Smith JL, Fischetti RF, Yamamoto M (2012) Micro-crystallography comes of age. Curr Opin Struct Biol 22(5):602–612
9. Ballesteros J, Weinstein H (1995) Integrated methods for modeling G-protein coupled receptors. Method Neurosci 25:366–428
10. Manglik A et al (2012) Crystal structure of the mu-opioid receptor bound to a morphinan antagonist. Nature 485(7398):321–326
11. Granier S et al (2012) Structure of the delta-opioid receptor bound to naltrindole. Nature 485(7398):400–404
12. Gregory KJ, Sexton PM, Christopoulos A (2007) Allosteric modulation of muscarinic acetylcholine receptors. Curr Neuropharmacol 5(3):157–167
13. Redka DS, Pisterzi LF, Wells JW (2008) Binding of orthosteric ligands to the allosteric site of the M(2) muscarinic cholinergic receptor. Mol Pharmacol 74(3):834–843
14. Dror RO et al (2013) Structural basis for modulation of a G-protein-coupled receptor by allosteric drugs. Nature 503(7475):295–299
15. Nygaard R et al (2013) The dynamic process of beta(2)-adrenergic receptor activation. Cell 152(3):532–542
16. Rasmussen SG et al (2011) Structure of a nanobody-stabilized active state of the beta(2) adrenoceptor. Nature 469(7329):175–180
17. Ring AM et al (2013) Adrenaline-activated structure of the β2-adrenoceptor stabilized by an engineered nanobody. Nature 502(7472):575–5799
18. Kruse AC et al (2013) Activation and allosteric modulation of a muscarinic acetylcholine receptor. Nature 504(7478):101–106
19. Croy CH (2014) Characterization of the novel positive allosteric modulator, LY2119620, at the muscarinic M2 and M4 receptors. Mol Pharmacol 86(1):106–115
20. Conn PJ, Christopoulos A, Lindsley CW (2009) Allosteric modulators of GPCRs: a novel approach for the treatment of CNS disorders. Nat Rev Drug Discov 8(1):41–54
21. Kruse AC et al (2014) Muscarinic acetylcholine receptors: novel opportunities for drug development. Nat Rev Drug Discov 13(7):549–560

Chapter 3
Epigenetic Drug Discovery

Chun-wa Chung

Abstract The molecular aetiology of disease relies on a complex interplay between our underlying genetic predispositions and the more dynamic epigenetic signatures that govern how our DNA is transcribed. Whilst there is no opportunity to alter the genome with which we are born, there may be potential to understand, modulate and reset aberrant epigenetic motifs that lie at the root of dysfunction and thereby restore health. Small molecule inhibitors of epigenetic proteins such as the 'writers', 'erasers' and 'readers' of histone post-translational modifications provide chemical tools to assist our biological understanding and offer the prospect of epigenetic drugs that may give distinct and profound pharmacology in a number of complex diseases. This article introduces the proteins involved in histone epigenetic gene regulation and highlights the potential of diverse drug discovery approaches to deliver chemical tools and clinical candidates for Jumonji demethylase enzymes and bromodomain reader proteins.

3.1 Introduction to Epigenetics

3.1.1 What Is Epigenetics?

Epigenetics refers to heritable changes in gene expression, resulting in a change in phenotype, that does not involve alterations to the underlying DNA sequence. Epigenetic mechanisms include DNA methylation, production of non-coding RNA, and histone post-translational modifications (PTMs) [1]. Together, these elements contribute to our epigenome, which governs what, where and when our DNA is translated and adds context-dependency to our gene regulation processes [2]. Our epigenetic profile is also influenced by environmental factors such as diet and early

C.-w. Chung (✉)
GlaxoSmithKline Research & Development, Gunnelswood Road, Stevenage, Hertfordshire SG1 2NY, UK
e-mail: chun-wa.h.chung@gsk.com

© Springer Science+Business Media Dordrecht 2015
G. Scapin et al. (eds.), *Multifaceted Roles of Crystallography in Modern Drug Discovery*, NATO Science for Peace and Security Series A: Chemistry and Biology, DOI 10.1007/978-94-017-9719-1_3

27

life experiences [3, 4]. These lectures will focus on one aspect of the epigenetic machinery, proteins involved in reading, writing and erasing the "histone code" of PTMs [5].

3.1.2 Interest in Epigenetic Proteins as Therapeutic Targets?

Unlike the DNA-coded genome, the epigenetic state of our cells is dynamic. Normally, it evolves through defined states during a cell's natural life cycle of differentiation and development in a tightly regulated manner. However, aberrant changes in our epigenome can also occur and there is increasing evidence that these may be important for the onset and maintenance of complex diseases such as cancer [6, 7], neurological disorders [8], inflammation [9] and metabolic illnesses [10]. Drug molecules that are able to target the critical epigenetic processes responsible for preserving these dysfunctional transcriptional states therefore offer the prospect of providing longer-lasting, as well as more effective treatment, through the simultaneous control of multiple genes [11–13].

3.1.3 Histone Readers, Writers and Erasers

Genomic material within the nucleus is stored compactly in a form called chromatin. This consists of tightly wound DNA wrapped around a histone octamer, containing one tetramer of histones H3 and H4 (two copies each) and two histone H2A–H2B dimers. Covalent PTMs of the exposed histone protein tails influence chromatin architecture and affect the accessibility of the DNA to transcription. There is an extensive list of possible histone modifications, but within the lectures we will focus on the two most widely studied histone modifications, acetylation [14] and methylation [15] (Fig. 3.1).

The complex pattern of histone PTMs, or "histone code", governs processes from simple gene expression to cell fate determination and, in some cases, disease onset and persistence. Only a subset of the possible PTM patterns are found physiologically, because the enzymes that put on (epigenetic writers) and remove (epigenetic erasers) these covalent marks do so in a sequence and context dependent manner [16]. The deciphering of this PTM code uses the synergistic action of a number of reader domains [17]. Often, each reader is designed to bind a specific mark rather weakly. However, when brought together, either as domains within the same protein or as part of different proteins within the megaDalton protein–protein complexes associated with chromatin, the multivalent interaction of the reader proteins confers specificity and affinity for the specific pattern of modification recognised [18] (Table 3.1).

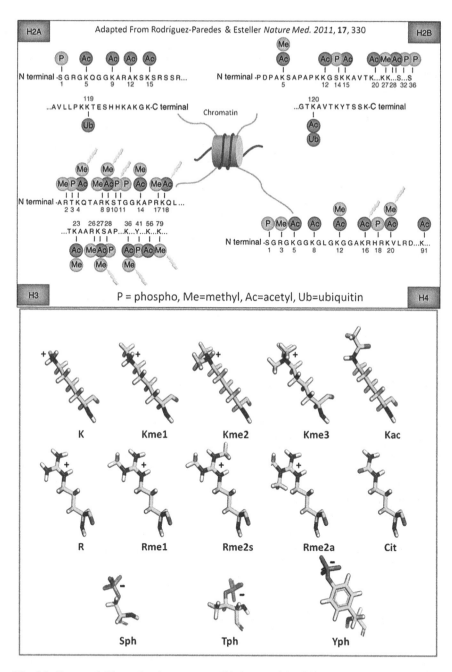

Fig. 3.1 *Top panel*: Figure showing some possible known sites of PTMs on histone tails. *Bottom panel*: Structures of common post-translational modifications of histone lysine and arginine residues and the standard abbreviation used to designate these. Row 1 shows lysine (K), its sequential methylation states (Kme) and acetylated lysine (Kac). Row 2 shows arginine (R), its possible sequential methylation states (Rme) and citrulline (Cit), which is produced by deimination of arginine. Row 3 shows phosphorylation of serine, threonine and tyrosine

Table 3.1 Readers, writers and erasers of histone code

Family	No. of proteins	Major classes
Readers		
Bromodomains (BD)	61	Major readers of acetylated-lysines
Methyl lysine or methyl arginine readers other than PHDs	95	Royal family: Tudor, Agenet, chromo, PWWP, MBT WD40r
PHD domains	104	Large and diverse family that reader unmodified and methylation and acetylated residues
Writers		
Histone acetyltransferases	18	MYST family, GNAT, EP300
Protein methyltransferases	60	SET domain: methylates lysines
		PRMT: methylates arginines
		PRDMs: SET domain-like
Erasers		
Histone deacetylases	17	Class I, II, IV: Zn-dependent
		Class III (Sirtuins): NAD-dependent
Lysine demethylases	25	Lysine Specific Demethylase 1, 2: flavin-dependent
		Jumonji: 2-oxoglutarate de

3.2 Chemical Probes for Jumonji Histone Demethylases

A lot remains unknown about the molecular mechanism by which histone readers, writers and erasers contribute to epigenetic-based human diseases. One way to increase our knowledge and understanding is to use chemical probes [19] that inhibit a known protein, or subset of proteins, involved in histone modification and to study the biological consequences of inhibition. Whilst this is a powerful strategy, the current lack of "quality" chemical probes with good affinity, defined selectivity and cell permeability often hinders the wider use of this methodology [20].

The case study of Jumonji histone enzymes illustrates the use of target-based approaches, including structure-based drug design (SBDD), to find chemical probes that reveal novel pharmacology [21].

3.2.1 What Are Jumonji Enzymes?

Jumonji (Jmj) enzymes are the largest class of histone lysine demethylases (KDM) [22]. These proteins regulate gene expression both by providing a vital 'scaffolding' role within transcriptional chromatin complexes, and by virtue of their enzymatic demethylase activity. The relative importance of these two roles and the therapeutic

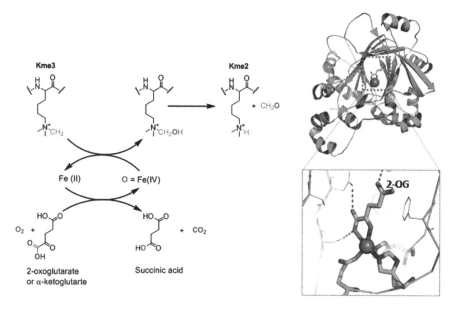

Fig. 3.2 Demethylation mechanism for Jumonji enzymes and structure of JMJD2A (2GP5.pdb) with insert showing active site

potential of Jmj proteins as drug targets have been much debated due to the lack of specific small molecule inhibitors to demonstrate chemical tractability and pharmacological relevance.

All Jumonji enzymes have a catalytic (JmjC) domain of the cupin fold that contains a conserved motif consisting of two histidines and either a glutamate or aspartate residue. These key residues coordinate the catalytic Fe^{2+} ion and the essential 2-OG co-factor (Fig. 3.2). In addition, many Jumonji enzymes possess one or more histone reader domains (e.g. PHD, Tudor, Fbox) that have been shown to be important for substrate specificity and localisation of the enzymes to their site of action [23] (Fig. 3.3).

3.2.2 Discovery and Optimisation of Chemical Probes for the KDM6 Subfamily

The KDM6 subfamily consists of three members: JMJD3, UTX and UTY. JMJD3 (KDM6B) is the most-well studied and functions as a specific demethylase of lysine-27 of histone H3 (H3K27). Studies have placed JMJD3 at key cell fate decision points in T lymphocytes and macrophages. For example, in macrophages, JMJD3 is rapidly induced through an NF-kB-dependent mechanism in response to bacterial

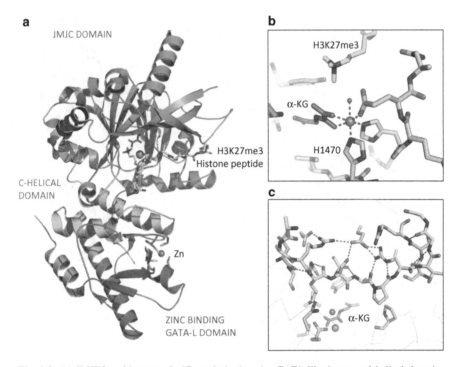

Fig. 3.3 (**a**) JMJD3 architecture: JmjC catalytic domain, GATA-like insert and helical domain. a-KG is shown in stick format, peptide in stick format and metal as a sphere (4EZH.pdb). Detailed views of metal and peptide binding are shown in (**b**) and (**c**)

products and inflammatory stimuli. JMJD3 has also been demonstrated to regulate the differentiation state of the epidermis, to activate the tumour suppressor INK4A-Arf5, and to be upregulated in prostate cancer.

A diversity screen against JMJD3 at GlaxoSmithKline (GSK) identified a series of weakly active hits which was optimised, using SBDD (Fig. 3.4), to yield the lead GSK-J1 (90 nM IC_{50} in JMJD3 ALPHA screen)

Liganded crystal structures of JMJD3 provided key insights into the specificity determinants for cofactor, substrate and inhibitor recognition. For example, unusually, GSK-J1 moves the catalytic metal out of its normal site (comparison of Figs. 3.3b and 3.4) so the metal ion no longer interacts directly with H1470 but does so via a bridging water molecule. This enables two nitrogens of the 2-(pyridine-2-yl)pyrimidine moiety of GSK-J1 to chelate the ion.

This structural knowledge allowed us to easily design an inactive control compound, the pyridine regio-isomer GSK-J2, where the nitrogen in the pyridine ring has been moved to prevent such a bidentate metal interaction. Although good *in vitro* tools, GSK-J1 and GSK-J2 were not perfect chemical tools as their acidic nature results in poor cell permeability. This was overcome by employing a pro-drug

ACTIVE CHEMICAL PROBES   INACTIVE CHEMICAL PROBES

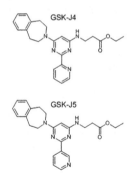

Fig. 3.4 Bound structure of GSK-J1 in JMJD3 (PDB code: 4ASK) and chemical structures of active (GSK-J1, GSK-J4) and inactive (GSK-J2, GSK-J5) tool compounds

strategy using ethyl ester derivatives to produce the cell penetrant compounds GSK-J4 and GSK-J5. These molecules are readily hydrolysed by intracellular esterases to give the parent acids. GSK-J4 and GSK-J5 therefore provide an ideal matched pair of active and inactive tool compounds to elucidate the functional role of JMJD3 inhibition.

3.2.3 Chemical Probes Reveal Role of KDM6 Catalytic Activity in Inflammation

Although JMJD3 is known to mediate an inflammatory response in cells exposed to lipopolysaccharide (LPS), the mechanism of this activity was not understood. Using GSK-J4 and GSK-J5, it was revealed that JMJD3 normally contributes to the LPS-induced inflammation response by removing the inhibitory H3K27me3 chromatin mark at the promoters of cytokine genes, such as TNF-α, thus allowing their transcription [21]. An inhibitor such as GSK-J4 may therefore have the potential to reduce inflammation in diseases such as rheumatoid arthritis. The dual inhibitory activity of GSK-J4 against JMJD3 and UTX was found to be crucial for effective TNF–α reduction, highlighting a previously unknown redundancy in the functions of these two enzymes.

This was the first example of a selective histone demethylase inhibitor, and demonstrates that small molecule inhibitors of histone-modifying enzymes have potential to improve our understanding of epigenetic regulation of gene expression and deliver novel drug classes to treat disease through epigenetic modulation.

3.3 Bromodomain Inhibitors from Phenotypic Hits to First-Time in Man

Phenotypic screening has been shown to be a powerful way of finding pharma-cologically innovative compounds. Recent analysis has shown this methodology lead to more approved first-in-class chemical entities (NCEs) than target-based approaches [24]. This is true both historically and since the change to target-focused drug discovery in the 1990s. However, subsequent identification of the molecular target and mechanism of action (MMOA) have been recognized as being very important in many instances. Knowledge of the MMOA facilitates a hypothesis-driven understanding of how first-in-class compounds may be advanced to best-in-class follower molecules with an improved safety profile or longer duration of action. It is notable that a number of epigenetic targets and inhibitors were initially discovered by phenotypic screening. These include the first HDAC inhibitor Vorinostat (SAHA) [25–27] and the first bromodomain inhibitors [28–31].

3.3.1 What Are Bromodomains?

Bromodomains (BDs) are small (~110 amino acid) evolutionarily and structurally conserved modules that bind acetyl-lysine. They are always found as part of much larger bromodomain containing proteins (BCPs). Many of which have roles in regulating gene transcription and/or chromatin remodelling. The human genome encodes for at least 56 bromodomains in 42 different proteins (Fig. 3.5) [32].

The three-dimensional structures of over half of the family have been determined by X-ray crystallography and nuclear magnetic resonance spectroscopy (NMR). These reveal that they share a common fold, consisting of four antiparallel α-helices (αZ-αA-αB-αC) arranged in a left-handed twist (Fig. 3.6). The acetyl-lysine binding site lies at one end of the helical bundle and the essential elements for recognition of this PTM are known to reside in key interactions with a conserved asparagine and tyrosine within this site. Specificity and affinity of cognate peptides is governed by two regions that flank this pocket, the ZA and BC loops. These regions differ widely amongst bromodomains and allow the histone context of the acetyl-lysine to be distinguished. As such they are also likely areas where inhibitors can exploit interactions to gain selectivity.

Unusually the binding site is lined with an extensive network of highly conserved water molecules (Fig. 3.6). For the most studied bromodomain family, the BET family, these waters are so constant that they can be effectively considered as part of the protein structure. H-bonding interactions between BET inhibitors and W1, the water that bridges to the conserved tyrosine, are found to be as important as those interactions to the conserved Asn [41] (Fig. 3.6).

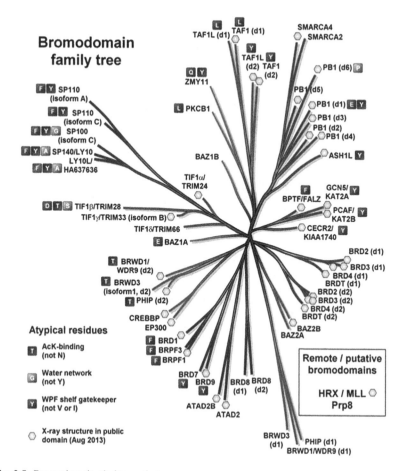

Fig. 3.5 Bromodomain phylogenetic tree

3.3.2 Discovery of BET Bromodomain Inhibitors by Phenotypic Screening

The BET (bromodomain and extra-terminal domain) family represents an important family of BCP drug targets. There are four family members: the ubiquitously expressed BRD2/3/4 and testis-specific BRDT. They have a conserved architecture consisting of N-terminal tandem bromodomains (BD1 and BD2) and a C-terminal ET (extra-terminal) domain.

The first potent bromodomain inhibitors disclosed, JQ1 [33] (**1**), I-BET762 and I-BET819 [34] (**2–3**) (Fig. 3.7) are pan-BET compounds that bind to the entire BET subfamily with nanomolar potency. They were derived from chemical starting points found by cellular assays focused on functional/phenotypic readouts [35].

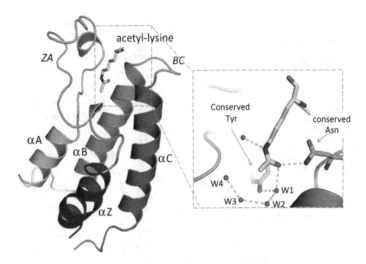

Fig. 3.6 The archetypal four-helix bundle topology of bromodomains (αZ-αA-αB-αC) is illustrated by the structure of BRD4-BD1 complexed to the H4K5acK8ac peptide (PDB: 3UVW). Highlighted at the *top* of the helical bundle are the ZA and BC loops that flank the acetyl-lysine binding pocket. Variability in these loops allows the AcK residues to be recognized in the context of differing peptide sequences. The insert on the *right* shows the conserved tyrosine and asparagine in stick format and the hydrogen-bonding network within the AcK binding site to the carbonyl of the acetyl

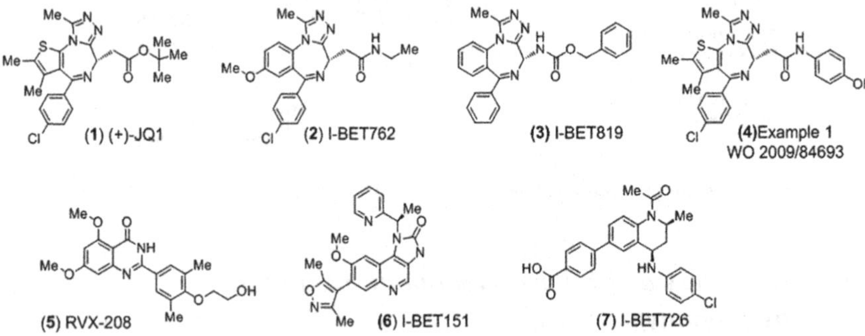

Fig. 3.7 Examples of BET bromodomain inhibitors

The GlaxoSmithKline (GSK) clinical candidate I-BET762 (**2**), and other GSK series (**6**, **7**), were initially derived from activators found in an Apo-A1 luciferase reporter assay. The molecular target for these compounds was deconvoluted using a combined chemoproteomic, biophysics and X-ray crystallographic approach. This revealed that these molecules were direct antagonists of acetyl-lysine histone binding to BET proteins [36] (Fig. 3.8).

Equally serendipitously, anti-inflammatory screening by Mitsubishi-Tanabe identified thienodiazepines (**4**) as potent inhibitors that were found to have BRD4

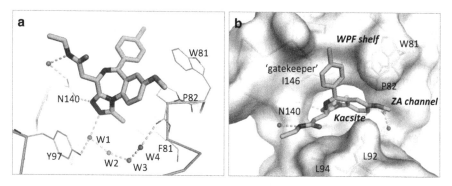

Fig. 3.8 (**a**) I-BET762 binding in BRD4-BD1 (PDB: 3P5O) (**b**) I-BET762 binding in BRD4-BD1 with protein surface depicted

activity and anti-proliferative effects [31, 36]. The way in which the molecular mode of action of the Mitsubishi compounds was unravelled has not been made public. However, their 2009 published patent was noticed by investigators, such as the Structural Genomics Consortium and their collaborators, and inspired the design of JQ1. Most recently, RVX-208 (**5**), a clinical candidate for atherosclerosis, acute coronary syndrome and Alzheimer's disease was also revealed to bind to the BET bromodomains [29, 37]. This compound was progressed as a regulator of ApoA1 gene transcription without a MMOA before the emerging BET literature prompted the testing of this compound for BET activity [38].

These BET compounds, which could be termed phenotypically derived chemical probes, have enabled many laboratories to contribute to our growing knowledge of BET function and further our understanding of epigenetic mechanisms in general. Proposed therapeutic opportunities for BET inhibitors include: oncology [12], male contraception [39], metabolic disease [40] and inflammation [9].

3.3.3 Beyond Phenotypic Screening

There are remarkably few small molecule tools to dissect the biology of bromodomain inhibition outside the BET family. There is thus a clear need for additional chemical probes to elucidate the specific biological consequence of bromodomain inhibition for this target class. The wealth of structural knowledge about this family of proteins suggests that a structure based drug design (SBDD) approach would be applicable to this target class [41–46]. Examples of this are now rapidly emerging within the literature, including those from the Structural Genomics Consortium (SGC) epigenetic probe discovery efforts that involve multiple academic and industrial partners [47]. Undoubtedly, many new chemical tools will be found by target-based approaches, however, difficulties in their use in unravelling pharmacology in this complex area should not be underestimated.

In summary, success in the discovery of epigenetic inhibitors has been achieved by a variety of diverse approaches. These chemical tools will be valuable in dissecting important biological questions [48]. A key challenge is the translation of those undoubtedly partial insights into delivery of clinically efficacious and differentiated medicines.

References

1. Goldberg AD, Allis CD, Bernstein E (2007) Epigenetics: a landscape takes shape. Cell 128:635–638
2. Gardner KE, Allis CD, Strahl BD (2011) Operating on chromatin a colorful language where context matters. J Mol Biol 409(1):36–46
3. Skinner MK, Manikkam M, Guerrero-Bosagna C (2010) Epigenetic transgenerational actions of environmental factors in disease etiology. Trends Endocrinol Metab 21(4):214–222
4. Jirtle RL, Skinner MK (2007) Environmental epigenomics and disease susceptibility. Nat Rev Genet 8(4):253–262
5. Strahl BD, Allis CD (2008) The language of covalent histone modifications. Nature 403(6765):41–45
6. Sharma S, Kelly TK, Jones PA (2010) Epigenetics in cancer. Carcinogenesis 31(1):27–36
7. Elsasser SJ, Allis CD, Lewis PW (2011) Cancer new epigenetic drivers of cancers. Science 331(6021):1145–1146
8. Urdinguio RG, Sanchez-Mut JV, Esteller M (2009) Epigenetic mechanisms in neurological diseases: genes syndromes and therapies. Lancet Neurol 8(11):1056–1072
9. Prinjha R, Tarakhovsky A (2013) Chromatin targeting drugs in cancer and immunity. Genes Dev 27(16):1731–1738
10. Gluckman PD, Hanson MA, Buklijas T, Low FM, Beedle AS (2009) Epigenetic mechanisms that underpin metabolic and cardiovascular diseases. Nat Rev Endocrinol 5(7):401–408
11. Egger G, Liang G, Aparicio A, Jones PA (2004) Epigenetics in human disease and prospects for epigenetic therapy. Nature 429(6990):457–463
12. Dawson MA, Kouzarides T (2012) Cancer epigenetics: from mechanism to therapy. Cell 150(1):12–27
13. Kelly TK, De Carvalho DD, Jones PA (2010) Epigenetic modifications as therapeutic targets. Nat Biotechnol 28(10):1069–1078
14. Kouzarides T (2000) Acetylation: a regulatory modification to rival phosphorylation? EMBO J 19(6):1176–1179
15. Kouzarides T (2002) Histone methylation in transcriptional control. Curr Opin Genet Dev 212(2):198–209
16. Berger SL (2007) The complex language of chromatin regulation during transcription. Nature 447(7143):407–412
17. Taverna SD, Li H, Ruthenburg AJ, Allis CD, Patel DJ (2007) How chromatin-binding modules interpret histone modifications: lessons from professional pocket pickers. Nat Struct Mol Biol 14(11):1025–1040
18. Ruthenburg AJ, Li H, Patel DJ, Allis CD (2007) Multivalent engagement of chromatin modifications by linked binding modules. Nat Rev Mol Cell Biol 8(12):983–994
19. Frye SV (2010) The art of the chemical probe. Nat Chem Biol 6(3):159–161
20. Allis CD, Muir TW (2011) Spreading chromatin into chemical biology. Chembiochem 12(2):264–279
21. Kruidenier L, Chung CW, Cheng Z, Liddle J, Che K, Joberty G, Bantscheff M, Bountra C, Bridges A, Diallo H, Eberhard D, Hutchinson S, Jones E, Katso R, Leveridge M, Mander PK, Mosley J, Ramirez-Molina C, Rowland P, Schofield CJ, Sheppard RJ, Smith JE, Swales C,

Tanner R, Thomas P, Tumber A, Drewes G, Oppermann U, Patel DJ, Lee K, Wilson DM (2012) A selective jumonji H3K27 demethylase inhibitor modulates the proinflammatory macrophage response. Nature 488(7411):404–408

22. Shi Y (2007) Histone lysine demethylases: emerging roles in development physiology and disease. Nat Rev Genet 8(11):829–833
23. Upadhyay AK, Horton JR, Zhang X, Cheng X (2011) Coordinated methyl-lysine erasure: structural and functional linkage of a Jumonji demethylase domain and a reader domain. Curr Opin Struct Biol 21(6):750–760
24. Swinney DC, Anthony J (2011) How were new medicines discovered? Nat Rev Drug Discov 10(7):507–519
25. Duvic M, Vu J (2007) Vorinostat: a new oral histone deacetylase inhibitor approved for cutaneous T-cell lymphoma. Expert Opin Investig Drugs 16(7):1111–1120
26. Richon VM, Webb Y, Merger R, Sheppard T, Jursic B, Ngo L, Civoli F, Breslow R, Rifkind RA, Marks PA (1996) Second generation hybrid polar compounds are potent inducers of transformed cell differentiation. Proc Natl Acad Sci U S A 93(12):5705–5708
27. Richon VM, Emiliani S, Verdin E, Webb Y, Breslow R, Rifkind RA, Marks PA (1995) A class of hybrid polar inducers of transformed cell differentiation inhibits histone deacetylases. Proc Natl Acad Sci U S A 95(6):3003–3007
28. Chung C, Coste H, White JH, Mirguet O, Wilde J, Gosmini RL, Delves C, Magny SM, Woodward R, Hughes SA, Boursier EV, Flynn H, Bouillot AM, Bamborough P, Brusq JM, Gellibert FJ, Jones EJ, Riou AM, Homes P, Martin SL, Uings IJ, Toum J, Clement CA, Boullay AB, Grimley RL, Blandel FM, Prinjha RK, Lee K, Kirilovsky J, Nicodeme E (2011) Discovery and characterization of small molecule inhibitors of the BET family bromodomains. J Med Chem 54(11):3827–3838
29. McLure KG, Gesner EM, Tsujikawa L, Kharenko OA, Attwell S, Campeau E, Wasiak S, Stein A, White A, Fontano E (2013) RVX-208 an inducer of ApoA-I in humans is a BET bromodomain antagonist. PLoS One 8(12):e83190
30. Adachi K et al (2006) Thienotriazolodiazepine compound and a medicinal use thereof International Patent No PCT/JP2006/310709 (WO/2006/129623): 2006
31. Miyoshi S, Ooike S, Iwata K, Hikawa H, Sugahara K (2009) Antitumor agent WO 2009084693, 9 July 2009
32. Chung CW (2012) Small molecule bromodomain inhibitors: extending the druggable genome. Prog Med Chem 51:1–55
33. Filippakopoulos P, Qi J, Picaud S, Shen Y, Smith WB, Fedorov O, Morse EM, Keates T, Hickman TT, Felletar I, Philpott M, Munro S, McKeown R, Wang Y, Christie AL, West N, Cameron MJ, Schwartz B, Heightman TD, La TN, French CA, Wiest O, Kung AL, Knapp S, Bradner JE (2010) Selective inhibition of BET bromodomains. Nature 468(7327):1067–1073
34. Nicodeme E, Jeffrey KL, Schaefer U, Beinke S, Dewell S, Chung CW, Chandwani R, Marazzi I, Wilson P, Coste H, White J, Kirilovsky J, Rice CM, Lora JM, Prinjha RK, Lee K, Tarakhovsky A (2010) Suppression of inflammation by a synthetic histone mimic. Nature 468(7327):1119–1123
35. Florence B, Faller DV (2001) You bet-cha: a novel family of transcriptional regulators. Front Biosci 6:D1008–D1018
36. Adachi K, Hikawa H, Hamada M, Endoh J-I, Ishibuchi S, Fujie N, Tanaka M, Sugahara K, Oshita K, Murata M (2011) Preparation of thienotriazolodiazepine compounds having an action of inhibiting the CD28 costimulatory signal in T cell. US Patent 8044042B2, 25 Oct 2011
37. Picaud S, Wells C, Felletar I, Brotherton D, Martin S, Savitsky P et al (2013) RVX-208, an inhibitor of BET transcriptional regulators with selectivity for the second bromodomain. Proc Natl Acad Sci U S A 110(49):19754–19759
38. McNeill E (2010) RVX-208 a stimulator of apolipoprotein AI gene expression for the treatment of cardiovascular diseases. Curr Opin Investig Drugs 11(3):357–364

39. Matzuk MM et al (2012) Small-molecule inhibition of BRDT for male contraception. Cell 150:673–684
40. Belkina AC, Denis GV (2012) BET domain co-regulators in obesity inflammation and cancer. Nat Rev Cancer 12(7):465–477
41. Chung CW, Dean AW, Woolven JM, Bamborough P (2012) Fragment-based discovery of bromodomain inhibitors part 1: inhibitor binding modes and implications for lead discovery. J Med Chem 55(2):576–586
42. Bamborough P, Diallo H, Goodacre JD, Gordon L, Lewis A, Seal JT, Wilson DM, Woodrow MD, Chung CW (2012) Fragment-based discovery of bromodomain inhibitors part 2: optimization of phenylisoxazole sulfonamides. J Med Chem 55(2):587–596
43. Hewings DS, Fedorov O, Filippakopoulos P, Martin S, Picaud S, Tumber A, Conway SJ (2013) Optimization of 3 5-dimethylisoxazole derivatives as potent bromodomain ligands. J Med Chem 56(8):3217–3227
44. Vidler LR, Filippakopoulos P, Fedorov O, Picaud S, Martin S, Tomsett M, Hoelder S (2013) Discovery of novel small-molecule inhibitors of BRD4 using structure-based virtual screening. J Med Chem 56(20):8073–8088
45. Picaud S, Da Costa D, Thanasopoulou A, Filippakopoulos P, Fish PV, Philpott M, Knapp S (2013) PFI-1 a highly selective protein interaction inhibitor targeting BET bromodomains. Cancer Res 73(11):3336–3346
46. Gehling VS et al (2013) Discovery design and optimization of isoxazole azepine BET inhibitors. ACS Med Chem Lett 4(9):835–840
47. Ferguson FM, Fedorov O, Chaikuad A, Philpott M, Muniz JR, Felletar I, Ciulli A (2013) Targeting low-druggability bromodomains: fragment based screening and inhibitor design against the BAZ2B bromodomain. J Med Chem 56(24):10183–10187
48. Bunnage ME, Chekler ELP, Jones LH (2013) Target validation using chemical probes. Nat Chem Biol 9(4):195–199

Chapter 4
Crystallography and Biopharmaceuticals

Richard Pauptit

Abstract Biopharmaceuticals generally describe drugs synthesized by biotechnology rather than chemistry, and are normally macromolecules such as proteins (vaccines, antibodies, hormones) or nucleic acids (RNA, DNA), but could also include synthetic biology ambitions such as designed therapeutic microorganisms.

AstaZeneca has an experimental protein crystallography research group at each of its three research sites (Sweden, USA and UK). These groups support small molecule drug design for local projects as their main focus. AstraZeneca acquired MedImmune in 2007 as a strategic boost of its biopharmaceutical capability. Currently, there is no experimental structure capability within MedImmune.

A question arose whether the current structural support facility could address MedImmune projects with business impact. This chapter illustrates two cases where structural support provided unique information that could be incorporated into the patenting strategy in a timely fashion. These are structures of antibody-antigen complexes, which provide definitive epitope characterization. The structure of two interleukins, IL-17A and IL-15 are presented in complex with neutralizing antibodies that have been subjected to in-vitro affinity maturation. For the IL-17A antibody, the affinity maturation optimization process resulted in seven amino changes increasing affinity 6-fold and activity 30-fold. For the IL-15 antibody, there were nine amino acid changes with affinity increasing 228-fold and activity increasing a staggering 40,000-fold. We were intrigued to see whether the structures would help explain this, as well as providing the definitive epitopes.

4.1 Introduction

Crystallography at AstraZeneca has traditionally involved small molecule drug design, supporting medicinal chemistry, by illustrating exactly how lead compounds are binding to their targets. This would give rise to design ideas and hypothesis-driven work, and a drug discovery project could be supported by a large number of

R. Pauptit (✉)
Discovery Sciences, AstraZeneca, Alderley Park, Cheshire SK104TG, UK
e-mail: richard.a.pauptit@gmail.com

© Springer Science+Business Media Dordrecht 2015
G. Scapin et al. (eds.), *Multifaceted Roles of Crystallography in Modern Drug Discovery*, NATO Science for Peace and Security Series A: Chemistry and Biology, DOI 10.1007/978-94-017-9719-1_4

structure determinations. There lies the justification for the initial efforts of obtaining a structure: we consider the year or so it might take for protocol development (construct design, expression, protein production, crystallization, initial structure determination) to be an investment, while pay-off is provided by high-impact and timely iterative protein-inhibitor complex structures that drive medicinal chemistry and that can be turned around rapidly once the protocol is robust. This has been the nature of our support for 22 years now. Our CrysIS database (CRYStallographic Information System) holds over 5,000 complex structures covering more than 300 drug discovery projects.

While small molecule drugs are typically chemically synthesized, biopharmaceuticals generally refer to drugs created by biotechnology. They tend to be biological macromolecules, including proteins (antibodies, vaccines, hormones, etc.) and nucleic acids (DNA, RNA), but could also include synthetic biology ambitions such as designed therapeutic microorganisms. AstraZeneca strategically boosted its biopharmaceutical capability with the purchase of Cambridge Antibody Technology (UK) in 2006 and MedImmune (Gaithersburg, USA) in 2007 – both are now using the MedImmune branding.

Monoclonal antibody therapy is one biopharmaceutical application. It exploits the ability of antibodies to bind specifically and tightly to a single antigen. It can target extracellular or cell surface targets (although recently expression of "intrabodies" within the cell has been achieved using gene therapy [1]). The first approved therapeutic antibody (1986) was muromonab, a CD3-specific transplant rejection therapy [2]. At first, the monoclonal antibodies were murine, produced using hybridoma technology [3] (for which Kohler & Milstein were awarded the 1984 Nobel Prize for Medicine and Physiology). This sometimes gave rise to immunogenicity problems limiting therapeutic success. The development of recombinant DNA technology, transgenic mice and phage display allowed production of chimeric, humanized or fully human monoclonal antibodies and the therapeutic antibody market has increased dramatically. The Wikipedia entry for "monoclonal antibody therapy" has a brief overview and currently includes a table of products up to 2011. Therapeutic monoclonal antibody product names use the suffix –*mab*.

Affinity maturation is the immunological process where B-cells produce antibodies with increased affinity on successive exposure to antigen. This uses mutation (somatic hypermutation, with mutation rates about 1 M times higher than in cell lines outside the lymphoid system) and selection (B-cell progeny compete for resources including antigen; those with highest affinity are selected to survive). Typically this results in one or two mutations in each CDR in the antigen-binding region. Ideally, therapeutic antibodies would have picomolar affinities to enable target antigen saturation and to allow favorable dosing regimes. To achieve this, the mutation and selection processes of affinity maturation can be applied using biotechnology strategies. Random mutations can be introduced into the CDRs and selection methods such as phage display can be used to generate optimized antibodies by *in vitro* affinity maturation [4, 5].

Fig. 4.1 Schematic of an antibody

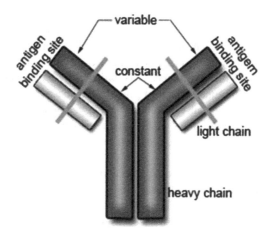

4.1.1   Basics of Antibody Structure

Antibodies, also known as immunoglobulins (Ig), come in various isoforms of which IgG is the most abundant. They are "Y" shaped molecules (Fig. 4.1) of about 150 KDa, with two heavy (H) and two light (L) chains. The protein chains are folded into a series of beta-sheet sandwiches known as immunoglobulin domains (four domains in the heavy chains, two in the light chains). The last immunoglobulin domain on each chain (on the tips of the "arms" of the Y) shows greater sequence variability and is known as the variable (V) domain, with the rest of the chain known as the constant (C) domain. At the tip of each variable domain, between the beta strands, there are the three "hypervariable loops" that form the antigen-binding site. These are also known as the three "Complementarity Determining Regions" (CDR1, CDR2, CDR3). On treatment with papain, the antibody is digested into three fragments corresponding to the stalk of the Y and the two arms of the Y. The stalk fragment is known as Fc (Fragment crystallizable) while the arm fragments are known as Fab (Fragment antigen-binding). Thus, a Fab fragment contains four immunoglobulin domains (C3H, C3L, VH, VL). VH and VL each contribute three CDR's, so that the antigen-binding surface (known as the paratope) may be composed of up to six hypervariable loops, allowing a very high degree of antigen specificity.

The most pertinent structural information that contributes to the intellectual property for therapeutic antibodies is a definitive mapping of the antigen epitope where the antibody binds: this may differentiate the candidate biopharmaceutical from competitor products. However, the X-ray crystal structure is but one way of obtaining epitope-mapping information, and it remains a matter of judgment whether additional epitope validation by X-ray crystallography justifies the effort.

It is clear that crystallography can and does provide key insights for biopharmaceutical development – for example, there is extensive work by Guy Dodson et al. on recombinant insulin, the first biopharmaceutical to be marketed in 1982. Therapeutic antibody engineering is a large field in itself [6]. The question we were facing was would the AstraZeneca crystallography capability be able to provide insights to MedImmune projects in a timely and hence impactful fashion, as for small molecule work? Typically, patenting strategy would aim to file rapidly after a candidate biopharmaceutical is identified, which can put pressure on the crystallography resource. In practice, success depends much on how rapidly suitable reagents may be obtained.

This chapter exemplifies two value-adding applications of crystallography to biopharmaceutical discovery, through crystal structure determination of antigen-Fab complexes. This work was an internal collaboration between MedImmune Cambridge and AstraZeneca and involved many contributors (see acknowledgements). The first example [7] is interleukin-17A in complex with a Fab fragment of the antibody CAT2200, where affinity maturation changed seven amino acid residues increasing affinity 6-fold (measured by SPR) and cellular activity 30-fold. The second example [8] is interleukin-15 in complex with the Fab fragment of DISC0280, where affinity maturation altered nine amino acid residues increasing the affinity 228 fold and the cellular activity a staggering 40,000 fold. We were intrigued to see whether the structures would help explain this.

4.2 Results and Discussion

4.2.1 Example 1: Interleukin 17A in Complex with CAT2200

4.2.1.1 Interleukin 17A and Disease

Interleukin-17A (IL-17A) is a secreted, glycosylated, homodimeric, pro-inflammatory cytokine produced by T-helper 17 (Th17) cells. It is the most studied member of the IL-17 cytokine family [9] (IL-17A to IL-17F) but no previous crystal structure was available, although a structure of the 50 % homologous IL-17F had been published [10]. The IL-17 cytokines mediate their effect by binding to the interleukin 17-receptor family of which there are five members [11] (IL-17RA to IL-17RE). IL-17A and IL-17F can bind either IL-17RA or IL-17RC, which may co-localize at the cell surface and function as heterodimeric receptors [12]. IL-17A is only found in small amounts in Th17 cells. It is expressed in disease compartments in a range of autoimmune diseases [13] including rheumatoid arthritis [14–16], multiple sclerosis [17, 18], psoriasis [19] and inflammatory bowel disease [20], and hence is viewed as a potential therapeutic target. A neutralizing antibody would bind the cytokine and prevent uptake by the receptor.

4.2.1.2 Isolation of the Antibody CAT2200

IL-17A binding antibodies were isolated [7] from a large phage library displaying human single-chain variable fragments (scFv) by panning selections on recombinant human IL-17A. A panel of scFv from these selections was identified by their ability to neutralize the binding of recombinant IL-17A to purified receptor. These scFv were reformatted as full-length IgG1 molecules and tested in a functional cell assay. The most potent lead antibody identified from the cell assay, TINA12, neutralized the activity of IL-17A with an IC50 of 23 nM.

Affinity maturation was carried out [7] separately for VH CDR3 and VL CDR3. CDR3 often contributes to specificity and it is known that this is an area where mutation can increase affinity [21]. scFv phage libraries containing CDR3 variants were subjected to multiple rounds of affinity-based phage display selections. A panel of optimized scFv was isolated from these selections through their improved ability to neutralize in the receptor-binding assay. These optimized scFv were reformatted as IgG1 and tested for neutralization in a cell assay. The VH and VL chains from several of the most potent antibodies were recombined, and the most potent recombined antibody was then reverted by mutagenesis to the closest human germline sequence producing CAT2200.

4.2.1.3 Structure Solution [7] of the Fab – Antigen Complex

The Fab fragment of CAT2200 was generated using papain and purified. Diffraction quality crystals grew in hanging-drop experiments after 2–3 weeks. Diffraction data were collected at the European Synchrotron Radiation Facility (ESRF), beamline ID29. The crystals are monoclinic $P2_1$ with cell dimensions a = 98.5 Å, b = 66.7 Å, c = 203.8 Å, and $\beta = 91.7°$. With two complex assemblies in the asymmetric unit, solvent content is 54 %, corresponding to a Matthews coefficient of 2.7 Å^3/Da.

The complex structure was solved by molecular replacement using the program MOLREP [22]. In a first trial, three molecules of the variable domain of Fab (generated from PDB 1AQK [23]) could be positioned. With these fixed in a second run, three molecules of the constant domain of Fab were found. The correctly generated Fab molecule indicated a correct MR solution, which then enabled location of the fourth Fab molecule by molecular replacement. The structure could be refined using REFMAC [22] and underwent manual rebuilding (in COOT [24]), giving R = 29.5 %. At this stage, molecular replacement failed to reveal the IL-17A dimer molecules, while density for these was visible in an F_o-F_c electron density difference map. Polyalanine strands could be built into $2F_o$-F_c electron density, eventually allowing use of a tryptophan residue as marker to identify a conserved pentapeptide, at which stage a model of IL-17F [10] could be overlaid. Now it was realized that a third of the IL-17A homodimer was disordered, and by adjusting the trail model accordingly, we found molecular replacement could be used successfully to place the second IL-17A homodimer. Structure refinement was completed to R = 22.6 %.

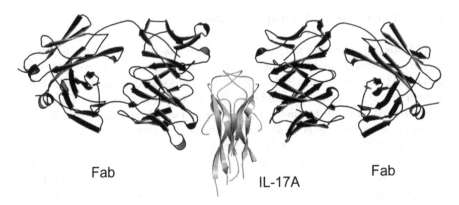

Fig. 4.2 Structure of the IL-17A – Fab complex

4.2.1.4   What Does the Structure Tell Us?

The first thing we see is that the IL-17A homodimer is sandwiched between two Fab fragments (Fig. 4.2). There are two equivalent IL-17A-Fab interaction sites, related by the IL-17A dimer symmetry. Then secondly, we can now see that the previously unknown structure of IL-17A is very similar to that of the closest related homolog IL-17F [10] (sequence identity 50 %). Overlay shows that the lower portion (approximately a third) of IL-17A is disordered. This does not appear to be influenced by crystal packing, lattice contacts are mediated through the Fab only. Like IL-17F, the structure adopts a cysteine knot architecture [10], though, as in IL-17F, one of the three classic disulfide bonds in this fold is missing: for the missing bond, the would-be knot Cys residues are instead Ser. From reducing and non-reducing SDS-PAGE indicating monomer and dimer, it is clear there must be an additional intermolecular disulfide, as is the case in IL-17F, but in IL-17A this falls in the disordered region. The IL-17F structure [10] has been described as a "garment", with a body, sleeves, collar and skirt. It is the skirt, which is disordered in IL-17A, while it is clear the packing allows ample room for the protein to be present. The collar and sleeves are the interaction sites with the antibody.

Thirdly, the crystal structure does indeed allow the epitope interactions between IL-17A and Fab to be examined in atomic detail and tabulated (Fig. 4.3). It is only necessary to describe one of the two interaction sites, since they are equivalent. The interactions involve all the complementarity determining regions (CDRs) from both the heavy(H) and the light(L) chain of the antibody fragment, and amino acid residues from both monomers (A and B) in the IL-17A dimer. The antibody heavy chain interacts with both chains A and B in the IL-17A dimer, while the light chain interacts only with one. In IL-17A, 12 amino acids form the epitope site, interacting with 16 amino acid residues in the antibody. The buried surface area per interface is around 760 $\mathring{A}^2$. The interactions include nine hydrogen bonds and numerous

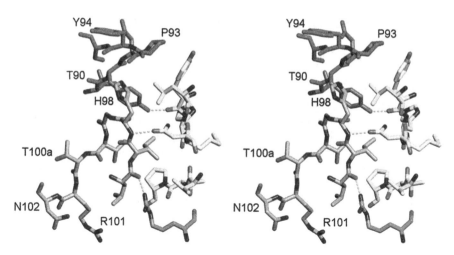

Fig. 4.3 Stereo diagram of the interface. Labeled residues have changed on affinity maturation. Residues from IL-17A homodimer are on the *right*, VL *top left*, VH *bottom left*

non-polar, van der Waals interactions. We can look at the affinity maturation mutants to try and understand their effect: only two of the seven affinity maturation mutants occur at the interface, and only in one can a stabilizing effect (stacking of a Pro93 ring) be inferred from the structure – suggesting there are multiple subtle effects that contribute to increased affinity.

Fourthly, we can examine the structure to understand how the antibody neutralizes the cytokine. The interface covers the equivalent of a central region in IL-17F, proposed at the time to be the receptor binding region, which contains a cavity in IL-17F, proposed to be a receptor binding pocket. No such cavity exists in the current IL-17A structure: a peptide forming one wall of the cavity is displaced relative to the IL-17F structure, filling the cavity. Thus the antibody may block receptor interaction (1) by sterically covering the receptor-binding site and (2) by inducing a conformational change, which affects the receptor-binding pocket. The displaced peptide is at the N-terminus of the ordered part of the IL-17A structure: it may even be that the disorder is induced by antibody binding. It would be of interest to examine a crystal structure of IL-17A in isolation. This was recently achieved together with an IL-17A – receptor structure [25], which indicates the flexibility, is functional: there is conformational rearrangement on receptor binding which is an allosteric effect that encourages heterodimeric receptor binding.

Fifthly, IL-17A is a homodimeric glycoprotein with a single glycosylation site at Asn45, although the recombinant protein used in these experiments was not glycosylated. The structure reveals that Asn45 is adjacent to the epitope, but is oriented toward solvent away from the antibody, suggesting the antibody would still be able to bind the glycosylated protein.

4.2.2   Example 2: IL-15 in Complex with DISC0280

4.2.2.1   IL-15 and Disease

Interleukin (IL) 15 is a member of the IL-2 family of cytokines that plays a key role in the activation and proliferation of natural killer cells and CD4+ T cells, and in the proliferation and maintenance of CD8+ T cells involved in memory responses to antigens [26, 27]. IL-15 binds to a trimeric receptor complex composed of its cognate receptor IL-15Rα, as well as β and γc chains that are shared with IL-2 [28]. Given the central role that IL-15 and its receptor play in the maintenance and persistence of adaptive immune response, IL-15 has been implicated as a key mediator in autoimmune diseases such as rheumatoid arthritis [29].

4.2.2.2   Isolation of the Antibody DISC0280

Again, the CDR3s were targeted for affinity maturation as these are often central to the binding interface [21]. However here [8], instead of two separate (VH and VL) libraries that could be recombined at a later stage to gain additional affinity as in the IL-17A example, five separate overlapping libraries were created each with six adjacent CDR residues that were varied: two of the libraries were for the VL CDR3 which has eleven residues (1–6; 6–11), and three of the libraries originate from the VH CDR3 which has sixteen residues (1–6; 6–11; 11–16). The libraries were subjected to three rounds of affinity-based solution-phase phage display selection with decreasing concentration of IL-15 at each round. The VH and VL libraries were pooled and individual variants were tested for IL-15 affinity in a competition-binding assay. Only variants from the first two VH libraries showed improved affinity. In contrast to the previous example, now the VL and VH pooled repertoires were recombined to form a single large library with mutations in VH and VL in order to explore any synergistic changes [8]. The recombined library underwent further affinity-based selection and individual ScFv were tested in the competition-binding assay, improved ScFv were converted to IgG1 and tested in a cell assay. The seven most potent of these all included mutations in VL, strongly suggesting synergy, since VL mutations alone gave no improvement pre-recombination. Similarly, the VH mutations in these most potent antibodies were now derived from the second and third sublibraries rather than the first two. DISC0280 was one of the optimized variants [8], showing a 40,000-fold improvement in cell-based activity over the parent antibody, with five mutations in VH CDR3 and four in VL CDR3.

4.2.2.3   Structure Solution of the DISC0280 Fab – Antigen Complex

The proteins were shown to bind with 1:1 stoichiometry by size exclusion chromatography. Crystals of monoclinic spacegroup C2 (a = 185.2, b = 43.8,

$c = 70.1$ Å, $\beta = 96.0°$) diffracted to 2.6 Å and diffraction data were collected on a Rigaku FRE generator. The structure [8] was solved by molecular replacement using program MOLREP [20]: again, correct generation of the Fab fragment after independent positioning of the variable and constant portions indicated a correct solution. On refinement of the Fab, difference density allowed modeling of the helices in IL-15 and the PDB structure could be easily superposed into the density and the complex structure was refined to $R = 24$ %.

4.2.2.4 What Does the Structure Tell Us?

The structure (Fig. 4.4) shows a 1:1 complex with the IL-15 cytokine adopting the same 4-helical bundle conformation seen in the published IL-15/IL-15R receptor complexes [30, 31]. The structure provides the desired definitive characterization of the epitope and paratope: 21 amino acid residues from the cytokine form the epitope and interact with 15 amino acid residues in the antibody paratope, giving an extensive and intimate interface with a buried surface area of 1,485 Å^2. This surface overlaps with the receptor-binding surface, hence the structure clearly indicates that neutralization by binding DISC0280 would prevent receptor binding.

Remarkably (Fig. 4.4), a key feature of the paratope is a 6-residue helix in VH CDR3. We believe this represents the first observation of secondary structure in a CDR. The helix contains all five affinity maturation mutants in VH CDR3, starting with the Glu100Pro mutation as helix initiator. The one helical residue that is not mutated (Gln100b) has intimate interactions with the antigen and is maintained during optimization. Also remarkably, the four mutations in VL CDR3 occur remotely from the interface, but are positioned immediately behind the helix

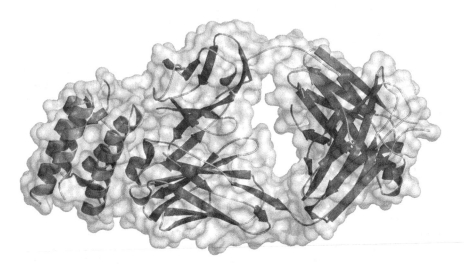

Fig. 4.4 Structure of IL-15 (*left*) complexed with DISC0280 Fab (*right*)

and would fine-tune the helix positioning. Thus the structure offers a rationale for the synergistic affinity mutations observed in VL CDR3. With the caveat that we do not have the structure of the parent antibody, assuming the helix is a consequence of mutation, it is tempting to suggest from these structural results that the stabilization of a pre-formed rigid binding surface offers a significant entropic advantage to the energetics of binding.

4.3   Conclusions

There is general consensus that crystallography is a preferred approach for definitive epitope mapping. In the two examples presented, the structures could be achieved rapidly, provided clear definition of the epitope, and could be readily incorporated into patenting strategy. This is not always necessarily the case. Opportunities can be missed typically because structure determination remains unpredictable and can fail to meet stringent time lines. It may be possible to focus on those cases where crystallization-quality reagents have already become available and time lines appear feasible. There may be additional opportunities for structural biology to contribute to scientific understanding beyond epitope characterization, e.g., rationalization of increased affinity.

The Structural Basis of Pharmacology: Deeper Understanding of Drug Design through Crystallography was a wonderful and successful meeting at Erice Sicily in the Spring of 2014 during which we celebrated a 40th anniversary of the course. It also marked my retirement after exactly 40 years of crystallography. I wish everyone peace, joy and continued enlightenment through crystallography.

Acknowledgements Many contributors enabled this work. Mark Abbott, Rick Davies, Maurice Needham, Melanie Snow, Caroline Langham, Wendy Barker, Adid Aziz, Sarah Dawson, and Malcolm Anderson at AstraZeneca prepared and characterized reagents; Fraser Welsch, Trevor Wilkinson, Tris Vaugan, Gerald Beste, Sarah Bishop, Bojana Popovich, Gareth Rees, Matthew Sleeman, Caroline Russel, David Lowe, Franco Ferraro, Debbie Patterson, Catriona Buchanon, Donna Finch and Phillip Mallinder at MedImmune Cambridge isolated and characterized the antibodies; Stefan Gerhardt and David Hargreaves at AstraZeneca did the crystallography; Steven Coales, Yoshitomo Hamura, and Steven Tuske at ExSAR provided additional epitope mapping by H/D exchange MS.

References

1. Chen SY et al (1994) Intracellular antibodies as a new class of therapeutic molecules for gene therapy. Hum Gene Ther 5(5):595–601
2. Hooks MA et al (1991) Muromonab CD-3: a review of its pharmacology, pharmacokinetics, and clinical use in transplantation. Pharmacotherapy 11(1):26–37
3. Milstein C (1999) The hybridoma revolution: an offshoot of basic research. Bioessays 21(11):966–973

4. Di Noia JM, Neuberger MS (2007) Molecular mechanisms of antibody somatic hypermutation. Annu Rev Biochem 76:1–22

5. Vaughan TJ et al (1996) Human antibodies with sub-nanomolar affinities isolated from a large non-immunized phage display library. Nat Biotechnol 14:309–314

6. Strohl WR, Strohl LM (2012) Therapeutic antibody engineering: current and future advances driving the strongest growth area in the pharmaceutical industry. Woodhead Publishing, ISBN-10: 1907568379; ISBN-13: 978-1907568374

7. Gerhardt S et al (2009) Structure of IL-17A in complex with a potent, fully human neutralizing antibody. J Mol Biol 394:905–921

8. Lowe DC et al (2011) Engineering a high-affinity anti-IL-15 antibody: crystal structure reveals an α-helix in VH CDR3 as key component of paratope. J Mol Biol 406:160–175

9. Rouvier E et al (1993) CTLA-8, cloned from an activated T cell, bearing AU-rich messenger RNA instability sequences, and homologous to a herpesvirus saimiri gene. J Immunol 150:5445–5456

10. Hymowitz SG et al (2001) IL-17s adopt a cystine knot fold: structure and activity of a novel cytokine, IL-17F, and implications for receptor binding. EMBO J 20:5332–5341

11. Shen F, Gaffen S (2008) Structure-function relationships in the IL-17 receptor: implications for signal transduction and therapy. Cytokine 41:92–104

12. Miossec P (2007) Interleukin-17 in fashion, at last: ten years after its description, its cellular source has been identified. Arthritis Rheum 56:2111–2115

13. Witowski J et al (2004) Interleukin-17: a mediator of inflammatory responses. Cell Mol Life Sci 61:567–579

14. Bessis N, Boissier MC (2001) Novel pro-inflammatory interleukins: potential therapeutic targets in rheumatoid arthritis. Joint Bone Spine 68:477–481

15. Nakae S et al (2003) Suppression of immune induction of collagen-induced arthritis in IL-17-deficient mice. J Immunol 171:6173–6177

16. Lubberts E et al (2004) Treatment with a neutralizing anti-murine interleukin-17 antibody after the onset of collagen-induced arthritis reduces joint inflammation, cartilage destruction, and bone erosion. Arthritis Rheum 50:650–659

17. Park H et al (2005) A distinct lineage of CD4 T cells regulates tissue inflammation by producing interleukin 17. Nat Immunol 6:1133–1141

18. Langrish CL et al (2005) IL-23 drives a pathogenic T cell population that induces autoimmune inflammation. J Exp Med 201:233–240

19. Arican O et al (2005) Serum levels of TNF-alpha, IFN-gamma, IL-6, IL-8, IL-12, IL-17, and IL-18 in patients with active psoriasis and correlation with disease severity. Mediators Inflamm 2005:273–279

20. Fujino S et al (2003) Increased expression of interleukin 17 in inflammatory bowel disease. Gut 52:65–70

21. Furukawa K et al (2001) A role of the third complementarity determining region in the affinity maturation of an antibody. J Biol Chem 276:27622–27628

22. Potterton E et al (2003) A graphical user interface to the CCP4 program suite. Acta Crystallogr D Biol Crystallogr 59:1131–1137

23. Faber C et al (1998) Three-dimensional structure of a human Fab with high affinity for tetanus toxoid. Immunotechnology 3:253–270

24. Emsley P, Cowtan K (2004) Coot: model-building tools for molecular graphics. Acta Crystallogr D Biol Crystallogr 60:2126–2132

25. Liu S et al (2013) Crystal structures of interleukin 17A and its complex with IL-17 receptor A. Nat Commun 4:1888. doi:10.1038/ncomms2880

26. Waldmann TA, Tagaya Y (1999) The multifaceted regulation of interleukin-15 expression and the role of this cytokine in NK cell differentiation and host response to intracellular pathogens. Annu Rev Immunol 17:19–49

27. Ma A et al (2006) Diverse functions of IL-2, IL-15, and IL-7 in lymphoid homeostasis. Annu Rev Immunol 24:657–679

28. Giri JG et al (1994) Utilization of the beta and gamma chains of the IL-2 receptor by the novel cytokine IL-15. EMBO J 13:2822–2830
29. Fehniger TA, Caligiuri MA (2001) Interleukin 15: biology and relevance to human disease. Blood 97:14–32
30. Chirifu M et al (2007) Crystal structure of the IL-15–IL-15Ralpha complex, a cytokine-receptor unit presented in trans. Nat Immunol 8:1001–1007
31. Olsen SK et al (2007) Crystal structure of the interleukin-15: interleukin-15 receptor α complex: insights into trans and cis presentation. J Biol Chem 282:37191–37204

Chapter 5
Structural Chemistry and Molecular Modeling in the Design of DPP4 Inhibitors

Giovanna Scapin

Abstract Inhibition of dipeptidyl peptidase IV (DPP-4) is an established approach for the treatment of type 2 diabetes. In 2006, Sitagliptin phosphate, a potent, orally bioavailable and highly selective small molecule DPP-4 inhibitor was approved by the FDA as once daily novel drug for the treatment of type 2 diabetes. Given the clinical success of sitagliptin our laboratories have been interested in generating analogues amenable for once-weekly dosing, to increase medication adherence. The first of such compounds was approved for preclinical and clinical development in 2008. During the back-up development stages, structural chemistry was used to generate new ideas, as well as evaluate in-silico proposals and screening results, and used to guide and significantly accelerate the drug discovery process.

5.1 Introduction

Inhibition of dipeptidyl peptidase IV (DPP-4) is an established approach for the treatment of type 2 diabetes. DPP-4 is the enzyme responsible for inactivating the incretin hormones glucagon-like peptide 1 (GLP-1) and glucose dependent insulinotropic polypeptide (GIP), two hormones that play important roles in glucose homeostasis [1–3]. DPP-4 inhibition increases circulating GLP-1 and GIP levels in humans, which lead to decreased blood glucose levels, hemoglobin A1c levels, and glucagon levels (for reviews of available data, see references [4–6]). In 2006, Sitagliptin phosphate, a potent, orally bioavailable and highly selective small molecule DPP-4 inhibitor [7, 8] was approved by the FDA as novel drug for the treatment of type 2 diabetes. Before approval, while sitagliptin was progressing through clinical trials, its structure, bound to DPP4 [7] together with the structures of a second class of inhibitors [9], was used as starting point for the structure-driven compound design used in the preparation of different back-up molecules [10–12]. These molecules were to retain or improve on sitagliptin properties and address some of its potential liabilities. Selectivity was one of the properties that needed to

---

G. Scapin (✉)
Structural Chemistry Department, Merck & Co., Inc, Kenilworth, NJ 07033, USA
e-mail: giovanna.scapin@merck.com

© Springer Science+Business Media Dordrecht 2015
G. Scapin et al. (eds.), *Multifaceted Roles of Crystallography in Modern Drug
Discovery*, NATO Science for Peace and Security Series A: Chemistry and Biology,
DOI 10.1007/978-94-017-9719-1_5

be maintained. DPP-4 is a member of a larger family of "DPP-4 activity- and/or structure-homologues" (DASH) enzymes, characterized by the common cleavage of a peptide bond found after a proline [13]. Except for DPP-IV, the functions of the DASH enzymes are mostly unknown, but given the common mechanism, they are likely to be involved in some of the biological processes that are regulated by proline-specific amino-terminal processes [14]. Thus, potential for off-target toxicity remained an issue that needed to be addressed in the development of suitable back-up compounds. This point was addressed by utilizing information from a combination of internal and external sources: modeling [15], 3D-QSAR approaches [11] and enzymes structure/activity analysis [16–18]. This resulted in several back-ups molecules being synthesized and tested, up to Sitagliptin approval.

After Sitagliptin approval, with no need to develop any more back-up compounds with properties similar to sitagliptin, the team shifted the focus to development a best-in-class DPP4 inhibitor with a once weekly dosing regimen. This was done to address some of the low adherence issues that manifest during the treatment of chronic diseases: the convenience of an effective, well-tolerated, weekly oral anti-hyperglycemic agent could have the potential to improve medication adherence, which may translate into better outcomes for patients with type 2 diabetes. The first of such compounds was approved for preclinical and clinical development in 2008 (MK-3102; [19, 20]), and soon after a new back-up program was initiated. The goal was to identify a long acting molecule with an in vitro and in vivo profile comparable or better to sitagliptin and MK-3102, and structurally different from both sitagliptin and MK-3102. This article will describe how structural chemistry was used throughout the back-up development stages and how a rational process, performed by molecular modeling with the support of structural data can guide and accelerate drug discovery significantly.

5.2   Brief Overview of the Drug Discovery Process

In a very simplified way, the drug discovery process can be subdivided in six steps as shown in Fig. 5.1: target identification and validation, lead identification, lead optimization, early development, late development and life-cycle management.

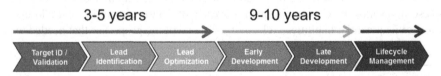

Fig. 5.1 Time line for drug discovery and development (Adapted from Ref. [21]). Developing a new drug, from target identification to product marketing, is a complicated and difficult process that can span many years. The cost of discovering, developing and marketing new drugs is over \$1 billion [21, 22], and it has been steadily increasing over the years [23]

The first three steps can be incorporated in the so called "early research and discovery (R&D) space". Although they account for about a third of the total process time [21], it has been long recognized the need to streamline the R&D process in order to substantially decrease the attrition rate and increase the probability of success for compounds entering clinical trials [24].

5.2.1   Target Identification/Validation

We have made great progress in curing many illnesses, both acute and chronic, that have been affecting the human race, but there are still many examples of unmet medical needs: cancer, diabetes, multi-resistant bacteria to name a few. The idea for a target (which is a term that can be applied to many diverse biological entities, from proteins to RNA) can come from a variety of sources, including academic and clinical research, and the commercial sector, and has been greatly helped by the advances in data mining, bioinformatics and phenotypic screening. Target validation may involve many different approaches: in vitro and in vivo experiments, genetics, tool compounds and literature analysis, to name a few [25]. The definition of unmet medical need does not refer only to diseases for which there is no cure, but also to the many cases in which a cure is no longer effective, or not sufficiently efficacious.

5.2.2   Lead Identification

There are several approaches that can be used to identify one or more small molecules with the desired inhibitory capabilities: high-throughput screening of large libraries is by far the most commonly used method for the initial identification of a lead compound, but focused libraries and fragment libraries have become more and more used as source of potential leads. The technologies used for lead identification have also evolved during the years, and while enzymatic and activity assays are commonly used in HTS screenings, biophysical techniques such as NMR, x-ray crystallography and surface plasmon resonance are often used with smaller fragment libraries [26]. Literature scouting, scaffold hopping, in-licensing, cheminformatics and computational techniques can also be productive ways to identify an initial compound [27].

5.2.3   Lead Optimization

There are many factors that need to be taken into account when developing a drug: potency is one, but stability, oral bioavailability, selectivity, good pharmacokinetics and in vivo activity against the desired target, among others, are all properties that need be obtained and/or improved during compound optimization. The identification of a development candidate is a collaborative effort between many groups. Compound optimization can take several years, and very often several molecules of the same of different classes are progressed to different stages and then abandoned before identifying a suitable candidate for clinical development [25].

5.2.4 Preclinical and Clinical Development

Early development includes preclinical developments as well as the first stages of the clinical studies. Preclinical development aims to provide a complete picture of how the compound behaves when it is administered to animals, as well as optimize production, formulation and delivery of the selected compounds. It is again a very collaborative effort between several groups of very diverse scientists, from safety assessment to drug metabolism to process chemistry. At the end of the pre-clinical development, an Investigational New Drug application is filed with the FDA, and, when approved, the IND triggers the compound entry in the Clinical development. This is by far the longest and most expensive portion of the drug discovery process. The first two phases of clinical trials aim to characterize the behavior of the compound in humans, providing information regarding absorption and metabolism, routes of elimination, safety, short-term side effects, and dose ranges. Phase III of the clinical trials aim to address safety and effectiveness in patients, as well less common and/or long term side effects. Labeling information and design are also established at this stage. It may take anywhere between 4 and 10 years for a new compound to move through the preclinical and clinical development stage, at the end of which the New Drug Application is filed [21, 28].

5.2.5 Lifecycle Management

Even after a compound is entered in the market, development work continues. Beside monitoring the use of the new compound in the larger population to ensure its long term safety and efficacy (or to uncover potential issues that were not evident in the limited sampling analyzed during clinical trials), novel applications of the compound are studied: these may be the addition of the new drug to an existing one to improve efficacy, or the discovery of a novel target, or the more accurate understanding of the drug mechanism of action. In any event, life cycle management extends for many years after a drug enters the market, and may provide much valuable information regarding not only the class of compounds but also the disease and the disease management.

5.3   DPP4 Inhibitor Back-Up Program

The back-up program was initiated with the goal of identifying a long acting molecule with an in vitro and in vivo profile comparable or better to sitagliptin and MK-3102, and structurally different from both sitagliptin and MK-3102. Multiple approaches were utilized in parallel to identify potential candidates: corporate library screening, structure-based drug design, and literature and patents based, modeling-guided scaffold hopping. Scaffold hopping, also known as lead hopping, is a strategy for generating novel chemical entities from other known drugs or compounds (For a recent review, see [29]). Scaffold-hopping strategies typically start from known active compounds and look to end with novel chemotypes: this can be achieved by different means (for example, opening and/or closing rings, swapping atoms (i.e., nitrogens and carbons in a heterocycle)), and may lead to different degrees of structural novelty depending on the chosen path. Scaffold hopping has been applied since the beginning of drug discovery, not only during lead identification but also in lead optimization approaches.

5.3.1   DPP4 Inhibitors Binding Site

The first structure of DPP4 in complex with a substrate analog was published in 2003 [30], and since then over 100 structures of DPP4s have been deposited in the Protein data Bank. DPP4 is a dimer of identical subunits. Each subunit contains an α/β hydrolase domain and an eight-bladed β-propeller domain. The substrate binding site is located in a smaller pocket within the large cavity formed by the two domains [30] which contains the serine protease active triad (Ser630, Asp708 and His740). A recent paper [31] proposed a common nomenclature for the identification of the different subsites located within the main binding site, due to the fact that, although no sites past S2 are formally defined in DPP4, most inhibitors bind well beyond S2 (Fig. 5.2). Based on this subdivision, inhibitors like Vildagliptin [31, 32] and saxagliptin [33, 34] are class 1 inhibitors, which occupy the S1 subsite (an hydrophobic pocket lined by Tyr662 and Tyr666 and containing the catalytic Ser630) and the S2 subsite, which contains the side chains of Tyr662, Glu205 and Glu206. Class 2 inhibitors, such as alogliptin [35] and linagliptin [36] occupy subsites S1, S1′ (identified by the side chain of Tyr547) and S2′ (containing Trp629). Sitagliptin is a class 3 inhibitor, and occupies subsites S1, S2 and an induced secondary binding pocket (S2 extensive subsite,) created by a flipping of the side chain of Arg358. This nomenclature will be used throughout the paper to describe the compounds binding mode.

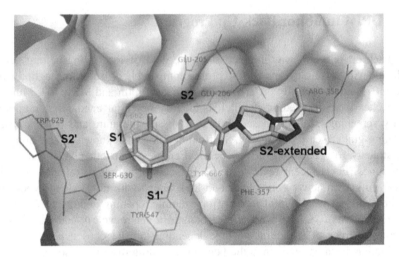

Fig. 5.2 Description of DPP4 binding site according to [31], using the structure of sitagliptin bound to DPP4 [7] as an example. The five subsites are labelled and the residues located within each subsite are identified. See text for a full description

5.3.2 MK-3102 Analogs

Several analogs were designed on the basis of MK-3102 structure to explore different ring sizes, various heteroatoms within the ring and different attachment points for the amino group. The goal was to retain the favorable interactions observed with MK-3102 [19] in subsites S1 and S2, while exploring different areas of the binding site (subsites S1′ and S2′). Figure 5.3a, b shows two of the proposed new molecules, and their predicted binding mode (based on the MK-3102 binding). For compound 1 the hypothesis was that substitution of a 6-member ring with a 4-member ring would retain the interactions within the S2 subsite, while projecting the pyrrolopyrazole moiety into the S1′ subsite, where it would interact with the side chain of Tyr547. For compound 2, the shifting of the amino groups was proposed to induce a 90-degree rotation of the central saturated ring, which would generate new possible positions for exploring the areas above and below the compound while retaining the stacking interactions of the pyrrolopyrazole with Phe357 in the S2 extensive subsite. Unfortunately the analogs that were synthesized showed a much poorer activity than expected. Figure 5.3c shows the binding mode identified by x-ray crystallography for two analogs. In both cases the binding mode was different than predicted and provided clue to the loss of potency: for compound 1, the pyrrolopyrazole moiety was positioned half-way between the S2 extensive and the S1′ subsites, essentially losing all interactions with the protein. For compound 2, the position of the thiane ring is, as expected, 90-degrees away from MK-3102 tetrahydropyran, but the pyrrolopyrazole is also rotated 90-degrees, with a substantial loss of the favorable interactions with Phe357. Based on the activity data and on the structural information work on this series of compound was quickly discontinued and resources allocated to other series.

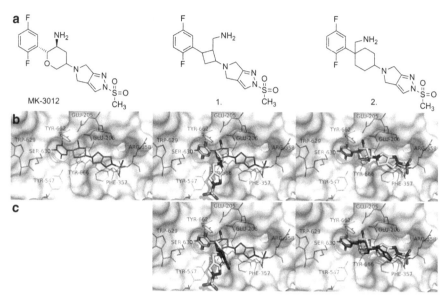

Fig. 5.3 (**a**) Chemical structure of MK-3102 [19] and of two of analogs proposed as part of the back-up program. (**b**) *Left panel* – the structure of an analog of MK-3102 [19] was used as template for the modeling driven design. *Center and right panel*: proposed binding mode for the two analogs depicted on top. MK-3102 is shown as *thin lines*, the two models as *thicker lines*. (**c**) Actual binding mode of analogs of the proposed compounds: MK-3102 is shown as *thin lines*, the two models as *thicker lines*, the actual structures as *thick darker lines*

5.3.3 Tricyclic Analogs

The very potent DPP4 inhibitor Linagliptin [36] contains a bicyclic xanthine core and has been shown to have a binding mode different from that of sitagliptin and MK-3102, with the xanthine moiety forming aromatic π-stacking interactions with the side chain of Tyr547 (Fig. 5.4, PDB entry 2RGU) in the S1′ subsite and the methylquinazoline stacked against the side chain of Trp629 in the S2′ binding pocket. Molecular modeling suggested the possibility of replacing the central xanthine with bio-isosteric cyclic guanines, thus providing a novel scaffold with possibly improved physical-chemical properties. Two replacements, cyclic guanines II and III (Fig. 5.5a) were proposed based on modeling. They differ for the point of attachment of the R_1 substituent but were predicted to be very similar in properties. The crystal structures of DPP-4 in complex with several compounds of this class were determined to atomic resolution and provided insights into the ligands binding mode and activity.

Both cyclic guanines II and III occupy the S1, S2, S1′ and S2′ subsites of the DPP-4 binding site [31], as shown in Fig. 5.5b. For both classes, the aminopiperidine nitrogen forms salt bridges with the conserved acidic residues (Glu205 and Glu206) and it is additionally hydrogen bonded to Tyr662 and one ordered water molecule.

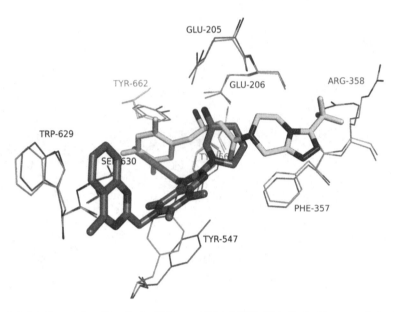

Fig. 5.4 Binding mode of linaglitpin (PDB entry 2RGU, [36]). Linagliptin (shown as *dark sticks*) extends into the S1′ and S2′ pockets, and interacts with the side chains of Tyr547 (pushing it into a different position from that observed in the sitagliptin (shown as *light sticks*) complex) and Trp629

The butynyl group extends into the largely hydrophobic S1 subsite. The tricyclic system is stacked against the side chain of Tyr547 in the S1′ prime site; the two aromatic systems are almost parallel to each other, and the average distance between them is ~3.6 Ang. The ring nitrogens are hydrogen bonded to water molecules, which are part of an extensive hydrogen-bonding network involving several waters as well as protein atoms. The benzyl of the methyl-quinazoline is parallel to and forms π-π stacking interactions with the side chain of Trp629 in the S2′ subsite. The distance between the two aromatic systems is ~3.6 Ang. The quinazoline nitrogens form hydrogen bonds to ordered water molecules that conversely interact with protein atoms. The methylene connecting the core tricyclic system to the methylquinazoline is also within hydrophobic interactions with the side chain of Tyr547. Despite the similar binding mode, representatives of guanines III were ~10-fold less potent than guanines II. A closer inspection of the structures reveals that the orientation of the compound with respect to the different side chains, and the relative distances are different in the two classes, and these differences may explain the observed differences in potency. Analysis of the structure of a scaffold III compound bound to DPP4 suggests that the drop in potency may be caused by the loss of positive hydrophobic interactions between the compound and the protein. Because of the different connection, the methylquinazoline is positioned farther away from the side chain of Trp629 and it is no longer parallel to Trp629. This looser interaction results in the methylquinazoline becoming disordered, and it has been found to assume two complete opposite orientations in the two binding sites

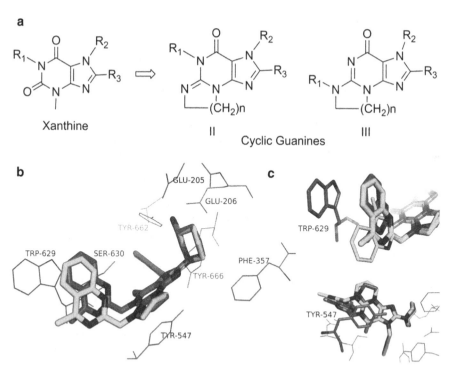

Fig. 5.5 (**a**) Xanthine replacements proposed on the bases of molecular modeling. (**b**) Both cyclic guanines II (*dark sticks*) and III (*light sticks*) bind as predicted, occupying subsites S1, S2, S1′ and S2′ [31] of the larger DPP4 binding site. (**c**) There are subtle differences in the binding mode of Guanine II and III (interactions with Trp629, *top panel* and Tyr 547, *bottom panel*) that may explain the difference in potency. See text for a full description

present in the structure: in one of the two orientation (thin light sticks in Fig. 5.5c, top panel), all interactions with Trp629 are lost; the other orientation (thick light sticks in Fig. 5.5c, top panel; this was the conformation predicted by modeling) is similar to the one observed in the scaffold II (thick, dark lines in Fig. 5.5c, top panel), but the quinazoline is farther away from the side chain of Trp629 than in scaffold II, and it is not properly oriented. In addition, the methylene linker between the tricyclic system and the quinazoline in scaffold III has been moved of about 2 Ang with respect to the position observed in scaffold II, and the hydrophobic interaction between the methylene linker and the side chain of Tyr547 are no longer possible (Fig. 5.5c, bottom panel).

Optimization of the R1 substituent on scaffold II quickly resulted in compounds with subnanomolar activity against DPP4. As observed for Linagliptin, binding of the R1 group to the S2′ site of DPP4, is responsible for selectivity over the related fibroblast activating protein (FAP), and SAR indicates that substitution of cyanobenzyl for methylquinazoline improves FAP selectivity up to 10,000-fold.

5.3.4 Alogliptin Analogs

Alogliptin is another potent and selective inhibitor of DPP-4 [35] which binds in the S1 and S1′ pockets of the DPP4 binding site (as shown in Fig. 5.7). Taking advantage of available sitagliptin and alogliptin X-ray crystal structures, and with the help of computer modeling, several series of heterocyclic compounds were designed as initial targets (Fig. 5.6). All compounds were thoroughly vetted through IP searches.

Unfortunately, most of the compounds showed reduced activity with respect to the parent compound. For example, the iminopyrimidones showed generally a 10–100 fold decrease in enzyme activity, and the crystal structure suggested that it could be possibly due to loss in π-πstacking interactions with Tyr547 and the proximity of one methyl group to the side chain of Ser630 (Fig. 5.7a). Compounds in which the nitrogen between the core ring and the S1 substituent was replaced by a carbon atom (C-linkers) also showed a moderate decrease in potency, although the overlay between alogliptin and the new compounds was very good (Fig. 5.7b).

Among all chemical classes derived from alogliptin, the spirocyclic derivatives represented a very unique and interesting group. During synthesis of the proposed spirocompound (Compound 3 in Fig. 5.8a), the opposite chirality was obtained (compound 4 in Fig. 5.8a), and the recovered compound was shown to have lost almost all the activity against DPP4 (the reported IC50 was greater than 10 μM). We were able to obtain a crystal structure of an analog of compound 4 bound to DPP4 (Fig. 5.8b) which revealed that the compound bound in a mode opposite to the one predicted for the original spirocyclic compounds, with the spirophenyl in S1 rather than in S1′, the cyano-benzene in S1′ and the and urea carbonyl interacting with Arg125. This binding mode was completely unexpected and induced some

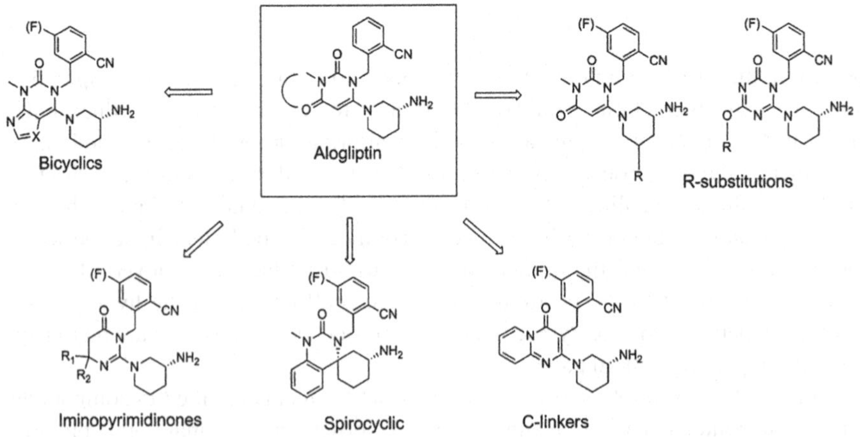

Fig. 5.6 Alogliptin [35] analogs proposed on the basis of the alogliptin and sitagliptin structures and computer modeling

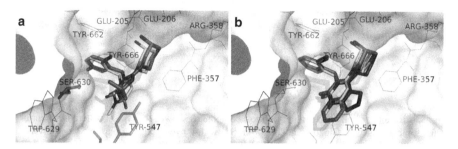

Fig. 5.7 (**a**) Binding of iminopyrimidinones (*dark sticks*) to DPP4, and comparison to alogliptin (*light sticks*). The π-staking interactions with Tyr547 in S1′ are lost, and one of the methyl groups is very close (3.3 Ang) to the side chain of Ser630. (**b**) Binding of two representatives of the C-linker compounds (*dark sticks*) and comparison to Alogliptin (*light sticks*). The overlay is very good, and the structure matches the predicted model, but the potency was somewhat decreased

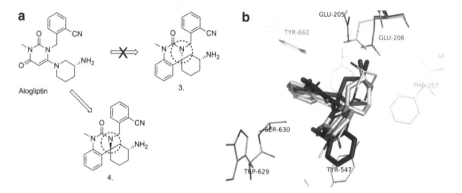

Fig. 5.8 (**a**) The structures of the desired, but not obtained, aloglitpin-derived spiro compound 3. (*top right*), and of the synthetically obtained spiro compound 4. (**b**) Predicted binding mode for the spiro-compounds (*light sticks*) and actual binding mode (*dark sticks*). In both cases the cyclohexylamine is bound within the S2 pockets, but the actual binding mode puts the spirophenyl in S1 rather than in S1′, and the cyano-benzene in S1′ instead of S1

movements within the binding site that were never seen before and that explain the loss of potency. The side chain of Glu206 moved from its normal position, and this resulted on sub-optimal salt-bridge interactions with the compound primary ammine. In addition, there was a clash between the urea N-methyl and the side chain of His740 in the S1 pocket. Some potency was probably recovered by a certain degree of pi stacking between the cyanobenzene and the side chain of Tyr547, but this was not enough. Nevertheless, this was a novel and unique binding mode, and several rounds of optimization were attempted to improve on the compound activity, but the resulting potency was always 10–20-fold less than that of sitagliptin.

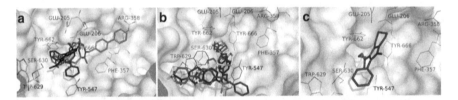

Fig. 5.9 Results of library screening: (**a**) most compounds showed the typical pattern of a central amine (interacting with Glu205 and Glu206 in S2) with a hydrophobic group in S1 and diverse substituents going into S1′ and S2′ or the S2 extended site. (**b**) Some compounds did not interact with residues in the S2 pockets, but extended into the S1′ and S2′ pockets. (**c**) This compound puts an ester (a polar substituent) in S1

5.3.5 Compound Libraries Screening

High Throughput Screening (HTS) of the Merck library identified almost a thousand potential hits, which were clustered according to structural similarity. Of the 153 clusters identified, 29 had more than 5 members. Several representative compounds from large size clusters, with potency less than 5 μM and an interesting or novel structural scaffold were requested for X-ray analysis. The results indicate a diverse sampling of the DPP4 binding pocket (Fig. 5.9), and some novel binding modes that were never previously reported. Most of the identified compounds retained the typical structure of a central amine (interacting with Glu205 and Glu206 in S2) with a hydrophobic group in S1 and diverse substituents going into S1′ and S2′ or the S2 extended site (Fig. 5.9a). Other compounds were identified that did not interact with the conserved Glu205/Glu206 in the S2 pocket, but occupied only the S1, S1′ and S2′ pockets, deriving most of the binding potency by hydrophobic interactions with the side chains of Tyr547 and Trp629 (similarly to linagliptin, Fig. 5.9b). In some cases (darker compounds in Fig. 5.9b) these compounds induced a conformational change in the position of Trp629, and wrapped around the side chain of the residue. The measured IC50 for these compounds was 1–2 μM, clearly indicating that the hydrogen bonds or salt bridges with the two glutamic acid residues in the S2 pocket are necessary for highly potent compounds, but provided a different framework onto which potency could be built. Another interesting example is shown in Fig. 5.9c: this compound interacts with Glu205 and Glu206, but puts a polar group in the mostly hydrophobic S1 pocket. In all, over 30 structures of novel hits were quickly made available to the chemistry team, thus providing an extensive chemical space that could be explored.

5.4 Conclusions

The MK-3102 back-up program only lasted about 1 year: MK-3102 was progressing smoothly through pre-clinical and clinical space, and the decision was made that a new back-up molecule was not necessary. Nevertheless, it provided a very good sampling of chemical space around the DPP4 binding pocket, and some very unusual and interesting results. The key component of the program was a very collaborative effort between Medicinal Chemistry, Structural Chemistry and Molecular Modeling, which quickly led to the discovery and/or design of potent and structurally distinct inhibitors. The program confirmed that a rational design process, including structure-based design, target based approaches and ligand-based approaches, performed by molecular modeling with the support of structural data, can accelerate drug discovery significantly, by providing assistance to the medicinal chemistry groups on selecting targets and prioritizing synthetic efforts.

Acknowledgments The author would like to thanks all the teams that worked on the DPP4 inhibitor programs over the years, especially the chemistry and biology leaders, Ann E Weber and Nancy A. Thornberry. Special thanks to Hubert Josien, Dmitri Pissarnitski and Wen Lian Wu, whose molecules are discussed in the paper. Use of the IMCA-CAT beamline 17-ID at the Advanced Photon Source was supported by the companies of the Industrial Macromolecular Crystallography Association through a contract with Hauptman-Woodward Medical Research Institute. Use of the Advanced Photon Source was supported by the U.S. Department of Energy, Office of Science, Office of Basic Energy Sciences, under Contract No. DE-AC02-06CH11357.

References

1. Mentlein R, Gallwitz B, Schmidt WE (1993) Dipeptidyl-peptidase IV hydrolyses gastric inhibitory polypeptide, glucagon-like peptide-1(7–36)amide, peptide histidine methionine and is responsible for their degradation in human serum. Eur J Biochem 214(3):829–835
2. Kieffer TJ, McIntosh CH, Pederson RA (1995) Degradation of glucose-dependent insulinotropic polypeptide and truncated glucagon-like peptide 1 in vitro and in vivo by dipeptidyl peptidase IV. Endocrinology 136(8):3585–3586
3. Deacon CF, Johnsen AH, Holst JJ (1995) Degradation of glucagon-like peptide-1 by human plasma in vitro yields an N-terminally truncated peptide that is a major endogenous metabolite in vivo. J Clin Endocrinol Metab 80(3):952–957
4. Deacon CF (2004) Therapeutic strategies based on glucagon-like peptide 1. Diabetes 53(9):2181–2189
5. Zerilli T, Pyon EY (2007) Sitagliptin phosphate: a DPP-4 inhibitor for the treatment of type 2 diabetes mellitus. Clin Ther 29(12):2614–2634
6. Deacon CF, Carr RD, Holst JJ (2008) DPP-4 inhibitor therapy: new directions in the treatment of type 2 diabetes. Front Biosci 13(5):1780–1794
7. Kim D, Wang L, Beconi M et al (2005) (2R)-4-oxo-4-[3-(trifluoromethyl)-5,6-dihydro[1,2,4]triazolo[4,3-a]pyrazin-7(8H)-yl]-1-(2,4,5-trifluorophenyl)butan-2-amine: a potent, orally active dipeptidyl peptidase IV inhibitor for the treatment of type 2 diabetes. J Med Chem 48(1):141–151
8. Thornberry NA, Weber AE (2007) Discovery of JANUVIA™ (Sitagliptin), a selective dipeptidyl peptidase IV inhibitor for the treatment of type2 diabetes. Curr Top Med Chem 7(6):557–568
9. Edmondson SD, Mastracchio A, Mathvink RJ et al (2006) (2S,3S)-3-amino-4-(3,3-difluoropyrrolidin-1-yl)-N, N-dimethyl-4-oxo-2-(4-[1,2,4]triazolo[1,5-a]-pyridin-6-ylphenyl) butanamide: a selective alpha-amino amide dipeptidyl peptidase IV inhibitor for the treatment of type 2 diabetes. J Med Chem 49(12):3614–3627
10. Biftu T, Scapin G, Singh S et al (2007) Rational design of a novel, potent, and orally bioavailable cyclohexylamine DPP-4 inhibitor by application of molecular modeling and X-ray crystallography of sitagliptin. Bioorg Med Chem Lett 17(12):3384–3387

11. Gao YD, Feng D, Sheridan RP et al (2007) Modeling assisted rational design of novel, potent, and selective pyrrolopyrimidine DPP-4 inhibitors. Bioorg Med Chem Lett 17(12):3877–3879
12. Edmondson SD, Mastracchio A, Cox JM et al (2009) Aminopiperidine-fused imidazoles as dipeptidyl peptidase-IV inhibitors. Bioorg Med Chem Lett 19(15):4097–4101
13. Rosenblum JS, Kozarich JW (2003) Prolyl peptidases: a serine protease subfamily with high potential for drug discovery. Curr Opin Chem Biol 7(4):496–504
14. Chen WT, Kelly T, Ghersi G (2003) DPPIV, seprase, and related serine peptidases in multiple cellular functions. Curr Top Dev Biol 54:207–232
15. Rummey C, Metz G (2007) Homology models of dipeptidyl peptidases 8 and 9 with a focus on loop predictions near the active site. PROTEINS: Struct Funct Bioinf 66(1):160–171
16. Kopcho LM, Kim YB, Wang A et al (2005) Probing prime substrate binding sites of human dipeptidyl peptidase-IV using competitive substrate approach. Arch Biochem Biophys 436(2):367–376
17. Bjelke JR, Christensen J, Branner S et al (2004) Tyrosine 547 constitutes an essential part of the catalytic mechanism of dipeptidyl peptidase IV. J Biol Chem 279(33):34691–34697
18. Longenecker KL, Stewart KD, Madar DJ et al (2006) Crystal structures of DPP-IV (CD26) from rat kidney exhibit flexible accommodation of peptidase-selective inhibitors. Biochemistry 45(24):7474–7482
19. Biftu T, Qian X, Chen P et al (2013) Novel tetrahydropyran analogs as dipeptidyl peptidase IV inhibitors: Profile of clinical candidate (2R,3S,5R)-2- (2,5-difluorophenyl)-5-(4,6-dihydropyrrolo [3,4-c]pyrazol-5-(1H)-yl)tetrahydro-2H-pyran-3-amine (23). Bioorg Med Chem Lett 19:5361–5366
20. Biftu T, Sinha-Roy R, Chen P et al (2014) Omarigliptin (MK-3102): a novel long-acting DPP-4 inhibitor for once-weekly treatment of type 2 diabetes. J Med Chem 57(8):3205–3212
21. DiMasi AJ, Hansen RW, Grabowski HG (2003) The price of innovation: new estimates of drug development costs. J Health Econ 22(2):151–185
22. DiMasi JA, Grabowski HG (2007) The cost of biopharmaceutical R&D: is biotech different? Manag Decis Econ 28(4–5):469–479
23. Morgan S, Grootendorst P, Lexchin J et al (2011) The cost of drug development: a systematic review. Health Policy 100(1):4–17
24. Kuhlmann J (1999) Alternative strategies in drug development: clinical pharmacological aspects. Int J Clin Pharmacol Ther 37(12):575–583
25. Hughes JP, Rees S, Kalindjian SB, Philpott KL (2011) Principles of early drug discovery. Br J Pharmacol 162(6):1239–1249
26. Kumar A, Voet A, Zhang KY (2012) Fragment based drug design: from experimental to computational approaches. Curr Med Chem 19(30):5128–5147
27. Guido RV, Oliva G, Andricopulo AD (2011) Modern drug discovery technologies: opportunities and challenges in lead discovery. Comb Chem High Throughput Screen 14(10):830–839
28. Ciociola AA, Cohen LB, Prasad Kulkarni P (2014) How drugs are developed and approved by the FDA: current process and future directions. Am J Gastroenterol 109(5):620–623
29. Sun H, Tawa G, Wallqvist A (2012) Classification of scaffold-hopping approaches. Drug Disc Today 17(7–8):310–324
30. Rasmussen HB, Branner S, Wiberg FC, Wagtmann NR (2003) Crystal structure of human dipeptidyl peptidase IV/CD26 in complex with a substrate analog. Nat Struct Biol 10:19–25
31. Nabeno M, Akahoshi F, Kishida H et al (2013) A comparative study of the binding modes of recently launched dipeptidyl peptidase IV inhibitors in the active site. Biochem Biophys Res Commun 434:191–196
32. Villhauer EB, Brinkman JA, Naderi GB et al (2003) 1-[[(3-hydroxy-1-adamantyl)amino]acetyl]-2-cyano-(S)-pyrrolidine: a potent, selective, and orally bioavailable dipeptidyl peptidase IV inhibitor with antihyperglycemic properties. J Med Chem 46(13):2774–2789
33. Augeri DJ, Robl JA, Betebenner DA et al (2005) Discovery and preclinical profile of Saxagliptin (BMS-477118): a highly potent, long-acting, orally active dipeptidyl peptidase IV inhibitor for the treatment of type 2 diabetes. J Med Chem 48(15):5025–5037

34. Metzler WJ, Yanchunas J, Weigelt J et al (2008) Involvement of DPP-IV catalytic residues in enzyme-saxagliptin complex formation. Protein Sci 17:240–250
35. Feng J, Zhang Z, Wallace MB et al (2007) Discovery of alogliptin: a potent, selective, bioavailable, and efficacious inhibitor of dipeptidyl peptidase IV. J Med Chem 50(10): 2297–2300
36. Eckhardt M, Langkopf E, Mark M et al (2007) 8-(3-(R)-aminopiperidin-1-yl)-7-but-2-ynyl-3-methyl-1-(4-methyl-quinazolin-2-ylmethyl)-3,7-dihydropurine-2,6-dione (BI 1356), a highly potent, selective, long-acting, and orally bioavailable DPP-4 inhibitor for the treatment of type 2 diabetes. J Med Chem 50:6450–6453

Chapter 6
Considerations for Structure-Based Drug Design Targeting HIV-1 Reverse Transcriptase

Eddy Arnold, Sergio E. Martinez, Joseph D. Bauman, and Kalyan Das

Abstract HIV-1 reverse transcriptase (RT) copies the viral single-stranded RNA genome into a double-stranded DNA version, and is a central target for anti-AIDS therapeutics. Eight nucleoside/nucleotide analogs (NRTIs) and five non-nucleoside inhibitors (NNRTIs) are approved HIV-1 drugs. Structures of RT have been determined in complexes with substrates and/or inhibitors, and the structures have revealed different conformational and functional states of the enzyme. Rilpivirine and etravirine, two NNRTI drugs with high potency against common resistant variants, were discovered and developed through a multidisciplinary structure-based drug design effort. The resilience of rilpivirine and etravirine to resistance mutations results from the structural flexibility and compactness of these drugs. Recent insights into mechanisms of inhibition by the allosteric NNRTIs include (i) dynamic sliding of RT/NNRTI complexes along template-primers and (ii) displacement of the RT primer grip that repositions the 3′-primer terminus away from the polymerase active site.

6.1 HIV-1 Drug Targets and Drug Resistance

AIDS started spreading silently in the 1970s, and came into the limelight in the early 1980s as a mysterious pandemic that was one of the most serious public health threats in history. Detection of reverse transcription activity in cultures of lymph node cells taken from AIDS patients in the early 1980s [1, 2] revealed that AIDS is caused by a retrovirus that was subsequently named the human immunodeficiency virus (HIV).

Reverse transcription is a key step in the life cycle of retroviruses, and the process is responsible for synthesis of double-stranded (ds) DNA from a viral single-stranded (ss) RNA genome. A viral DNA polymerase or reverse transcriptase (RT)

E. Arnold (✉) • S.E. Martinez • J.D. Bauman • K. Das
Center for Advanced Biotechnology and Medicine, Rutgers University, 679 Hoes Lane West, Piscataway, NJ 08854, USA
e-mail: arnold@cabm.rutgers.edu; kalyan@cabm.rutgers.edu

© Springer Science+Business Media Dordrecht 2015
G. Scapin et al. (eds.), *Multifaceted Roles of Crystallography in Modern Drug Discovery*, NATO Science for Peace and Security Series A: Chemistry and Biology, DOI 10.1007/978-94-017-9719-1_6

enzyme is responsible for synthesis of DNA complementary to an RNA or DNA template. Not surprisingly, the reverse transcription/DNA polymerization step in HIV replication was immediately considered a prime drug target, and the nucleoside analog AZT (zidovudine, ZDV) [3, 4] was approved as the first anti-AIDS drug in 1987. Clinical use of AZT revealed that treatment of HIV infection with a single drug was not effective in keeping the viral load down for a prolonged period. Also, it was realized that HIV could not be cleared from an infected individual, and drug-resistant viruses emerged with loss of sensitivity to AZT.

Even today, challenges like discovery of an effective AIDS vaccine, complete cure from HIV infection, and effective ways to overcome drug resistance remain as obstacles. However, coordinated research commitments initiated and supported by public and private initiatives have led to the development of drugs and treatment strategies that help HIV infected individuals lead a near normal lifespan if the patient complies with treatments to maintain viral load at or near undetected level. HIV-1 is the predominant virus that spreads AIDS. So far 28 individual anti-AIDS drugs have been approved, of which 13 target RT. The remaining drugs target other key steps in the viral lifecycle, namely: (i) the enzyme protease that is responsible for cleaving the viral polyprotein precursors into functional entities and for maturation of virus particles, (ii) the enzyme integrase that integrates the viral dsDNA to the host cell chromosome, and (iii) viral entry/fusion.

HIV-1 exhibits high genetic variability, and thereby, HIV-1 develops resistance to existing drugs and escapes host immune responses elicited by AIDS vaccine candidates. HIV-1 enters a host cell by binding the CD4 receptor on the surface of immune cells and a co-receptor, either CCR5 or CXCR4. Entry and fusion of an HIV-1 particle releases its two copies of the viral ssRNA genome, about fifty copies of RT, and other viral entities into the cytoplasm. RT copies the viral genome into a dsDNA that is subsequently transported into the nucleus of the infected cell and integrated into the host cell chromosome by another viral enzyme, integrase. Recently approved integrase inhibitors (raltegravir and elvitegravir) bind the active site of HIV-1 integrase in a pre-integration complex [5, 6] and block the viral DNA integration into host cell chromosomes.

Usually, three or four drugs are combined in a treatment regimen, commonly referred to as highly active antiretroviral therapy (HAART). Some of the widely used treatment regimens include (i) two nucleoside RT inhibitors (NRTIs) + one non-nucleoside RT inhibitor (NNRTI) or protease inhibitor (PI), and (ii) combinations of an integrase or entry inhibitor with RT inhibitors and PIs. Selecting the right treatment strategies and combinations remain challenging due to numerous factors. However, three decades of extensive research on various aspects of HIV-1 and related viruses have brought success in effectively managing HIV-1 infection, and the scientific knowledge garnered has in many ways blazed the trail for finding treatment solutions for several challenging chronic diseases and emerging drug-resistance problems.

6.2 Structures and Conformations of RT

RT is a heterodimer consisting of two polypeptide chains p66 (66 kDa; 560 amino acid residues) and p51 (51 kDa; 440 amino acid residues). Large precursor polyproteins translated from the HIV-1 pol gene are cleaved by HIV-1 protease to produce the p66 subunit. The p66 chain contains an N-terminal polymerase domain and a C-terminal RNase H domain. A likely scenario appears to be that two p66 chains dimerize, and HIV-1 protease cleaves the RNase H moiety of one chain to produce a stable p66/p51 heterodimeric RT. The structural architecture of RT (Fig. 6.1a) has been known for two decades [7, 8]. The p66 chain contains both the polymerase and RNase H active sites. The polymerase domain of RT (Fig. 6.1b) resembles the shape of a "right hand" with fingers, palm, and thumb subdomains; the connection subdomain links the polymerase domain to the RNase H domain. The p51 chain also contains the fingers, palm, thumb, and connection subdomains; however, the subdomains have different spatial arrangements in p51 than in p66. The subdomains in p51 are assembled into a relatively rigid structure that provides structural support to p66. The subdomains in p66 are flexible, and can rearrange to different conformational states necessary to carry out the functions of RT.

Structures representing five conformational/functional states of RT have been determined (Fig. 6.2). Some of the major conformational rearrangements revealed by RT structures are: (i) the thumb lifts up to bind nucleic acid, (ii) the fingers fold down to hold a dNTP substrate in the presence of a nucleic acid, (iii) NNRTI-binding leads to thumb hyperextension even if RT is not bound to a nucleic acid, and (iv) nucleic acid at the polymerase active site is repositioned upon binding of an NNRTI to RT-DNA complex. These structural rearrangements of RT result from inter-subdomain hinge movements and local structural rearrangements while the overall structural folds of individual subdomains remain almost invariant.

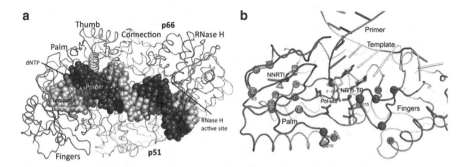

Fig. 6.1 HIV-1 RT structure and sites for common drug-resistance mutations [31]. (**a**) A ribbon representation of the structure of HIV-1 RT-DNA-dNTP complex. (**b**) Sites of commonly observed NRTI-resistance mutations (*magenta*) and NNRTI-resistance mutations (*cyan*) in the fingers and palm subdomains of HIV-1 RT (Color figure online)

72   E. Arnold et al.

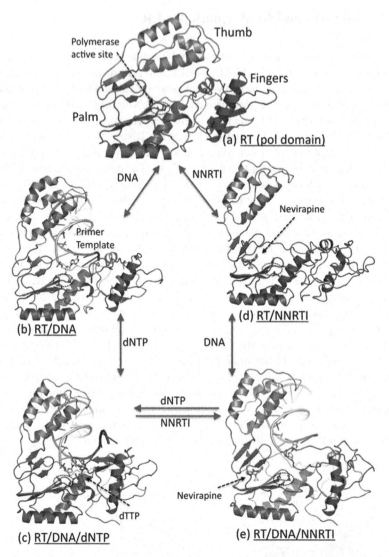

Fig. 6.2 Five structurally characterized conformational states of HIV-1 RT [31]; only the polymerase domain (fingers (*blue*), palm (*red*), and thumb (*green*)) of RT is shown. (**a**) Thumb is positioned near the fingers in a closed conformation that occupies the nucleic acid binding cleft in apo HIV-1 RT structures. (**b**) Upon binding of a nucleic acid substrate, the thumb is lifted up and acts as a clamp to hold the nucleic acid; the primer 3′-end is positioned at the polymerase active site (D110, D185, and D186) that is highlighted by a dotted ellipse. (**c**) Binding of a dNTP to RT-DNA complex results in closing of the fingers to bind a dNTP in a catalytically competent state. (**d**) Binding of an NNRTI to RT causes several conformational changes; thumb subdomain is lifted to a hyper-extended position and the nucleic acid-binding cleft is open even in absence of a nucleic acid. (**e**) Binding of an NNRTI to RT-DNA complex resulted in reduction of DNA interactions with the polymerase domain of RT, and repositioning of the primer 3′-end away from the polymerase active site; the structural study revealed no ordered dNTP binding to RT-DNA-NNRTI complex (Color figure online)

The sites of mutations that confer resistance to either NRTI or NNRTI drugs are primarily located in the polymerase domain (Fig. 6.1b). Routine RT sequencing from clinical isolates were limited to the polymerase domain only where primary NRTI- and NNRTI-resistance mutations occur. Relatively recent sequencing of the complete RT from clinical isolates have shown that mutations in the remote connection subdomain and RNase H domain enhance resistance to both classes of RT drugs [9, 10] by indirect mechanisms that are not well understood.

6.3 Nucleoside RT Inhibitors (NRTIs)

Nucleotide misincorporations by RT contribute to generating mutant HIV-1 proteins including mutant RTs, and mutant RTs can develop resistance to RT inhibitors. Drugs targeting HIV-1 RT are either nucleoside RT inhibitors (NRTIs) or non-nucleoside RT inhibitors (NNRTIs). An NRTI (Fig. 6.3) is converted to a dNTP analog by a phosphorylation cascade performed by cellular kinases, and then RT catalytically incorporates the drug as an NRTI monophosphate (or a nucleotide analog) at the 3′-end of the growing viral DNA primer; pyrophosphate is released as the reaction byproduct. The nucleoside phosphonate analogs like tenofovir (a "nucleotide analog") require addition of β−and γ−phosphates whereas the other NRTIs are elaborated with α−, β−, and γ−phosphates. Efficient intracellular phosphorylation of NRTI to NRTI-triphosphate (TP) is an essential requirement for the efficacy of an NRTI drug. Upon incorporation, an NRTI inhibits the elongation of DNA primer because NRTIs lack a 3′-OH group and/or contain a modified sugar moiety that prevents addition of the next nucleotide.

An NRTI-TP does not block the activity of an RT molecule; however, certain RT mutations cause NRTI resistance by discriminating an NRTI-TP from the analogous dNTP substrate. NRTI-resistance mutations primarily appear along the dNTP-binding track extending from the $\beta3 - \beta4$ fingers loop region to YMDD residue M184 (Fig. 6.1b). Also, RT has the ability to remove certain NRTIs from the DNA primer ("unblocking") by reversing the direction of its catalytic reaction of polymerization [11, 12]. The molecular mechanisms of individual

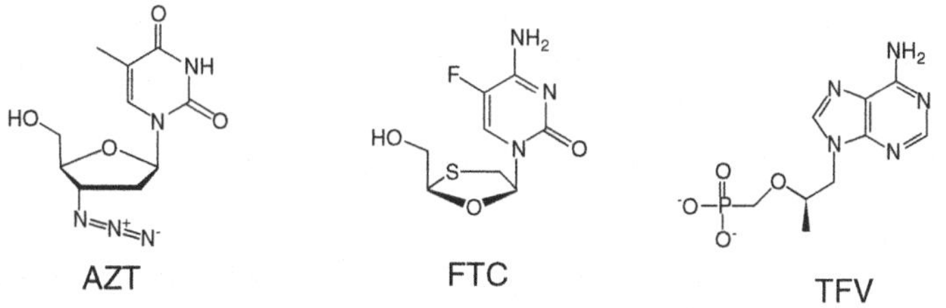

Fig. 6.3 Chemical structures of selected nucleoside RT inhibitors (NRTIs)

resistance mutations are unique and relationships among the mutations are complex. Biochemical and structural studies have illuminated some of the unique resistance mechanisms and their relationships.

6.4   Non-Nucleoside RT Inhibitors (NNRTIs)

Unlike NRTIs that do not directly inhibit RT, an NNRTI drug binds to a hydrophobic pocket in the palm subdomain adjacent to the base of the thumb subdomain (Fig. 6.2d) and allosterically inhibits DNA polymerization. The NNRTI pocket permits the design of highly specific inhibitors having low toxicities and minimal side effects. In fact, the NNRTIs are HIV-1 specific and even do not effectively inhibit HIV-2 RT. The NNRTI pocket is not required to be highly conserved for carrying out the enzyme activity unlike the conserved active site or dNTP-binding site of RT. Therefore, HIV-1 has a relatively lower genetic barrier for developing NNRTI-resistance mutations than for NRTI-resistance mutations. Primary NNRTI-resistance mutations appear in and around the NNRTI pocket, i.e., most of the pocket residues can mutate to confer NNRTI resistance. There are five NNRTI drugs approved for treating HIV-1 infections—of which nevirapine (NEV, Viramune®), efavirenz (EFV, Sustiva®), etravirine (ETR, Intelence®), and rilpivirine (RPV, Edurant®) are broadly used (Fig. 6.4a). An NNRTI is generally used in combination with NRTIs because the two classes of RT drugs have non-overlapping inhibition mechanisms and resistance mutation sites, and NNRTI-resistance mutations can emerge relatively quickly.

The first-generation NNRTI drug nevirapine invokes resistance mutations even after a single dose that is usually given to pregnant mothers to prevent mother-to-child transmission of HIV-1. An effective NNRTI should overcome the impacts of common drug-resistance mutations, a challenge that emerged with the discovery of first-generation NNRTIs in early 1990s. The initial structure of the RT-nevirapine complex [7] revealed the hydrophobic NNRTI-binding pocket which is ~10 Å away from the polymerase active site. The first-generation NNRTIs nevirapine, TIBO, and α-APA were found to assume a common "butterfly-like" binding mode despite their broad chemical diversity [13]. These NNRTIs could be optimized to nanomolar inhibitors of wild-type virus/enzyme; however, they can drastically lose their potency against a single common NNRTI-resistance mutation such as K103N or Y181C.

The diarylpyrimidine (DAPY) NNRTI drugs etravirine and rilpivirine were developed out of a structure-based multidisciplinary approach [14–16], and these drugs inhibited HIV-1 carrying common NNRTI-resistance mutations. The DAPY NNRTIs exhibit conformational flexibility that helps the drugs reorient "wiggle" and reposition "jiggle" to retain binding efficacy to RT even when pocket mutations emerge [14] (Fig. 6.4b). Further, the structures of rilpivirine in complexes with wild-type and two double mutant (K103N + Y181C and K103N + L100I) RTs demonstrated how the NNRTI rilpivirine wiggles and jiggles in the pocket to

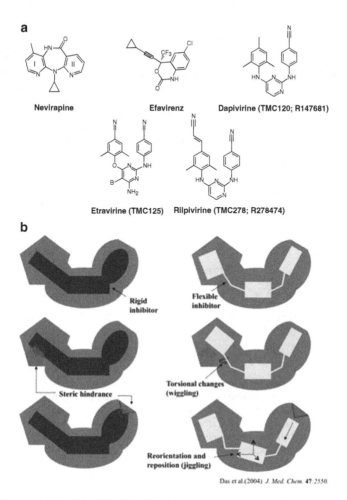

Fig. 6.4 Structural flexibility of NNRTIs [14]. (**a**) Chemical structures of selected non-nucleoside RT inhibitors (NNRTIs). (**b**) DAPY NNRTIs overcome common drug-resistance mutations by wiggling and jiggling

retain its binding affinity [17] (Fig. 6.5a–d). The predominant inhibitor-protein interactions associated with NNRTI binding are: (i) hydrophobic sandwiches, (ii) a characteristic hydrogen bond with the K101 main-chain carbonyl, and (iii) water-mediated hydrogen bonds. It is important that an NNRTI retains these interactions with RT while it wiggles and jiggles in NNRTI-pocket. The most recent NNRTI drug rilpivirine exhibits the flexibility to wiggle and jiggle by which it retains all of the above interactions including the key hydrogen-bonding interaction with the K101 main-chain carbonyl when it binds wild-type RT or the two double mutants [17]. A recent study shows that an NNRTI designed to have additional hydrogen bonds with RT while maintaining its flexibility has improved resilience against

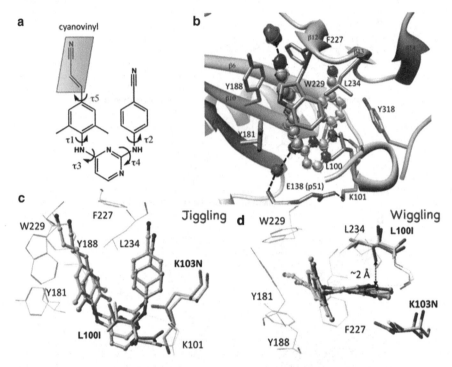

Fig. 6.5 Wiggling and jiggling of an NNRTI to retain potency against drug-resistance mutations [17] (**a**) Chemical structure of rilpivirine; the five torsionally flexible bonds define the conformational freedom of the NNRTI. (**b**) Thermal ellipsoid representation of rilpivirine drawn using the anisotropic B-factors of individual atoms from a 1.5 Å resolution structure of RT-rilpivirine complex (PDB ID 4G1Q). The interacting side chains and water molecules (*red*) are displayed. (**c** and **d**) Comparison of the binding modes of rilpivirine to wild-type RT (*gray*) vs. L100I + K103N mutant RT (*cyan*) revealed how the drug jiggles and wiggles, respectively, to evade the effects of drug-resistance mutations; the mutations modify the side chains from *yellow to orange* (Color figure online)

resistance mutations [18]. The cyanovinyl group of rilpivirine can swivel to maintain interactions with RT (Fig. 6.5a); addition of the cyanovinyl group, i.e., chemical modification from TMC120 → rilpivirine, enhanced the inhibition potency by ~3-fold. A recent study combining 2D infrared (IR) spectroscopy, high-resolution (1.51 Å) crystal structure, and molecular dynamics simulation of rilpivirine bound to RT demonstrated the involvement of the cyanovinyl group with an invariant water-mediated interaction that may be a critical feature for future NNRTI design considerations (Fig. 6.5b) [19]. Rilpivirine possesses highly favorable pharmacokinetics that, in combination with its high resilience to drug-resistance mutations, demonstrate clinical efficacy even at relatively low doses of 25 mg/day (efavirenz is usually administered at 600 mg/day). This ideal pharmacokinetic characteristic of rilpivirine may correlate with its property to form ordered nanoparticles with diameter ~100 nm, whereas several related compounds including etravirine which did not exhibit as favorable pharmacokinetics, were found to form disordered aggregates rather than uniform nanoparticles [20]. Understanding the molecular mechanism of NNRTI inhibition and resistance caused by different mutations may help with designing both future drugs and optimal drug combinations.

6.5 RT Inhibition by NNRTIs

RT inhibition by NNRTIs is indirect—NNRTIs are allosteric inhibitors. Early structures of RT and RT-NNRTI complexes showed that NNRTI binding traps RT in a rigid conformational state with an open nucleic acid-binding cleft; however, the impact of NNRTI on the of nucleic acid or dNTP, and above all, the mechanism of NNRTI inhibition of DNA polymerization remained elusive. Pre-steady-state and steady-state kinetic experiments in the mid-1990s suggested that the binding of an NNRTI inhibits the chemical step of DNA polymerization [21, 22]. A single-molecule FRET study revealed that RT frequently flips and slides over a double-stranded nucleic acid substrate [23, 24]; binding of dNTP stabilized the RT-DNA complex in a polymerase-competent mode while binding of an NNRTI induced destabilizing effects causing increased dissociation/association of RT with a double-stranded nucleic acid. The postulated mechanisms of inhibition by an NNRTI were: (i) restriction of thumb mobility [7], (ii) distortion of the catalytic triad [25] that would block the chemical step of DNA polymerization by RT [21], (iii) blockage of conformational state transition and not the chemical step leading to nucleotide incorporation [26], (iv) repositioning of the primer grip [27], and (v) loosening of the thumb and fingers clamp [23]. Multiple inhibition hypotheses evolved because NNRTI binding causes multiple structural and conformational changes in RT, and the direct and indirect contributions of individual changes towards the NNRTI inhibition were not clear.

The hinge motion between the thumb and palm is essential for the translocation of RT along the nucleic acid following each nucleotide incorporation. The breathing space that is transiently created traps an NNRTI. Upon binding of an NNRTI, RT loses its conformational mobility to carry out nucleotide addition and translocation. The NNRTI pocket does not exist in any RT structure that does not contain a bound NNRTI. A direct consequence of NNRTI binding is the opening of the NNRTI pocket to accommodate the inhibitor. The pocket formation requires the $\beta 12 - \beta 13 - \beta 14$ sheet, which contains the "primer grip", to move away from the $\beta 6 - \beta 10 - \beta 9$ sheet, which contains the catalytic triad (D110, D185, and D186). The recent structure of a ternary RT-DNA-nevirapine complex provides a snapshot of the effects of an NNRTI on DNA polymerization [28] (Fig. 6.6a, b). The binding of nevirapine shifts the primer grip which concomitantly displaces the primer 3'-end by ~5.5 Å away from its position at the polymerase active site. The interaction between the template-primer and the polymerase domain of RT is decreased upon nevirapine binding, which correlates with the earlier observation from a single-molecule study [23]. The fingers subdomain has an open conformation that would allow the entry of dNTPs (Fig. 6.6c); however, the repositioned template-primer

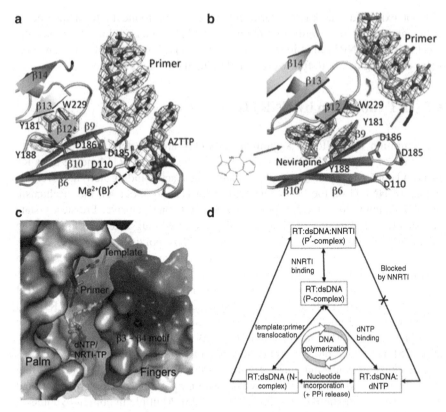

Fig. 6.6 Structural basis for the inhibition of DNA polymerization by an NNRTI [32]. (**a**) Structure of RT-DNA–AZTTP ternary complex obtained by soaking AZTTP into crystals of RT-DNA complex. (**b**) Soaking of nevirapine into the crystal created the NNRTI pocket, repositioning the 'primer grip' (on the β12–β13–β14 sheet) that moved the primer terminus away from the polymerase active site. (**c**) Electrostatic potential surface of RT bound to DNA and nevirapine. The crystallization experiments and structures showed an open dNTP-binding cleft into which dNTPs/NRTI-TPs can enter; however, structural perturbation by NNRTI binding did not allow a dNTP to chelate metals and form base-pairing and base-stacking interactions. (**d**) These structural constraints preclude the formation of an RT-DNA–dNTP polymerase competent (P) complex, rather forms a non-productive (P′) complex in the presence of an NNRTI—a structural basis for NNRTI inhibition

would not permit the base-pairing or base-stacking to support the binding of a dNTP at the N-site. This appears to be a primary reason why ordered binding of dNTP (or analog) to an RT-DNA complex was not observed in the presence of nevirapine whereas RT-DNA-dNTP (or analog) complexes could be formed when no NNRTI was present [28]. These experiments were carried out in crystals that permitted rearrangements of the polymerase domain upon binding of a dNTP/analog or an NNRTI. Our attempts to form an RT-DNA-nevirapine-AZTTP complex yielded the structure of only RT-DNA-nevirapine complex, and no well-defined binding of AZTTP at the polymerase active site when nevirapine is bound to RT [28].

In fact, a recent isothermal titration calorimetric (ITC) study revealed the inability of RT-DNA-NNRTI complex to bind a dNTP [29]. The structural information and published biochemical/biophysical results suggest that dNTPs may enter the dNTP-binding cleft (Fig. 6.6c), interact with RT, and may induce conformational changes of RT like closing of the fingers; however, formation of a catalytically relevant RT-DNA-dNTP complex would not be permitted when an NNRTI is bound. The repositioning of the template-primer by nevirapine binding disfavors base-pairing or base-stacking of a dNTP at the N site. In addition to base-pairing and base-stacking, metal chelation at the active site also contributes to the binding of a dNTP substrate in a catalytically competent RT-DNA-dNTP complex. None of the structures of RT-NNRTI binary complexes or the RT-DNA-nevirapine complex had any metal ion present at the polymerase active site, suggesting that a potential distortion of the catalytic site [25, 30] by an NNRTI also forbids ordered binding of the triphosphate moiety of a dNTP because of the loss of Mg^{2+} ion chelation at the active site. Thereby, NNRTI binding prevents RT from achieving a conformational state of RT-DNA (or RT-DNA/RNA) complexed with dNTP that would be required for catalysis [26]. Additionally, the structure of RT-DNA-nevirapine complex showed that (i) the thumb restriction [7] is induced by the primer grip repositioning, and (ii) the thumb and fingers clamp is loosened [23] by the reduced interactions between the polymerase domain of RT and DNA. All of these effects of NNRTI binding force RT into a structurally and catalytically non-competent complex for dNTP binding at the N-site and for polymerization (Fig. 6.6d).

Acknowledgments EA is grateful to the National Institutes of Health for support from grants R37 AI027690 (MERIT Award) and P50 GM103368. We also thank our collaborators in RT studies, both past and present.

Future Reading The reader may refer to the two reviews by Das and Arnold [31, 32] for further information. Much of the material and illustrations herein is reproduced from these two recent publications.

References

1. Barre-Sinoussi F et al (1983) Isolation of a T-lymphotropic retrovirus from a patient at risk for acquired immune deficiency syndrome (AIDS). Science 220:868–871
2. Gallo RC et al (1984) Frequent detection and isolation of cytopathic retroviruses (HTLV-III) from patients with AIDS and at risk for AIDS. Science 224:500–503
3. Mitsuya H et al (1985) 3′-Azido-3′-deoxythymidine (BW A509U): an antiviral agent that inhibits the infectivity and cytopathic effect of human T-lymphotropic virus type III/lymphadenopathy-associated virus in vitro. Proc Natl Acad Sci U S A 82:7096–7100
4. Fischl MA et al (1987) The efficacy of azidothymidine (AZT) in the treatment of patients with AIDS and AIDS-related complex. A double-blind, placebo-controlled trial. N Engl J Med 317:185–191
5. Grobler JA et al (2002) Diketo acid inhibitor mechanism and HIV-1 integrase: implications for metal binding in the active site of phosphotransferase enzymes. Proc Natl Acad Sci U S A 99:6661–6666

6. Hare S et al (2010) Molecular mechanisms of retroviral integrase inhibition and the evolution of viral resistance. Proc Natl Acad Sci U S A 107:20057–20062

7. Kohlstaedt LA, Wang J, Friedman JM, Rice PA, Steitz TA (1992) Crystal structure at 3.5 Å resolution of HIV-1 reverse transcriptase complexed with an inhibitor. Science 256:1783–1790

8. Jacobo-Molina A et al (1993) Crystal structure of human immunodeficiency virus type 1 reverse transcriptase complexed with double-stranded DNA at 3.0 Å resolution shows bent DNA. Proc Natl Acad Sci U S A 90:6320–6324

9. Nikolenko GN et al (2005) Mechanism for nucleoside analog-mediated abrogation of HIV-1 replication: balance between RNase H activity and nucleotide excision. Proc Natl Acad Sci U S A 102:2093–2098

10. Yap SH et al (2007) N348I in the connection domain of HIV-1 reverse transcriptase confers zidovudine and nevirapine resistance. PLoS Med 4:e335

11. Meyer PR, Matsuura SE, So AG, Scott WA (1998) Unblocking of chain-terminated primer by HIV-1 reverse transcriptase through a nucleotide-dependent mechanism. Proc Natl Acad Sci U S A 95:13471–13476

12. Arion D, Kaushik N, McCormick S, Borkow G, Parniak MA (1998) Phenotypic mechanism of HIV-1 resistance to 3′-azido-3′-deoxythymidine (AZT): increased polymerization processivity and enhanced sensitivity to pyrophosphate of the mutant viral reverse transcriptase. Biochemistry 37:15908–15917

13. Ding J et al (1995) Structure of HIV-1 RT/TIBO R 86183 complex reveals similarity in the binding of diverse nonnucleoside inhibitors. Nat Struct Biol 2:407–415

14. Das K et al (2004) Roles of conformational and positional adaptability in structure-based design of TMC125-R165335 (Etravirine) and related non-nucleoside reverse transcriptase inhibitors that are highly potent and effective against wild-type and drug-resistant HIV-1 variants. J Med Chem 47:2550–2560

15. Janssen PA et al (2005) In search of a novel anti-HIV drug: multidisciplinary coordination in the discovery of 4-[[4-[[4-[(1E)-2-cyanoethenyl]-2,6-dimethylphenyl]amino]-2-pyrimidinyl]amino]benzonitrile (R278474, rilpivirine). J Med Chem 48:1901–1909

16. de Bethune MP (2010) Non-nucleoside reverse transcriptase inhibitors (NNRTIs), their discovery, development, and use in the treatment of HIV-1 infection: a review of the last 20 years (1989–2009). Antiviral Res 85:75–90

17. Das K et al (2008) High-resolution structures of HIV-1 reverse transcriptase/TMC278 complexes: strategic flexibility explains potency against resistance mutations. Proc Natl Acad Sci U S A 105:1466–1471

18. Chong P et al (2012) Rational design of potent non-nucleoside inhibitors of HIV-1 reverse transcriptase. J Med Chem 55:10601–10609

19. Kuroda DG et al (2013) Snapshot of the equilibrium dynamics of a drug bound to HIV-1 reverse transcriptase. Nat Chem 5:174–181

20. Frenkel YV et al (2005) Concentration and pH dependent aggregation of hydrophobic drug molecules and relevance to oral bioavailability. J Med Chem 48:1974–1983

21. Spence RA, Kati WM, Anderson KS, Johnson KA (1995) Mechanism of inhibition of HIV-1 reverse transcriptase by nonnucleoside inhibitors. Science 267:988–993

22. Rittinger K, Divita G, Goody RS (1995) Human immunodeficiency virus reverse transcriptase substrate-induced conformational changes and the mechanism of inhibition by nonnucleoside inhibitors. Proc Natl Acad Sci U S A 92:8046–8049

23. Abbondanzieri EA et al (2008) Dynamic binding orientations direct activity of HIV reverse transcriptase. Nature 453:184–189

24. Liu S, Abbondanzieri EA, Rausch JW, Le Grice SF, Zhuang X (2008) Slide into action: dynamic shuttling of HIV reverse transcriptase on nucleic acid substrates. Science 322:1092–1097

25. Ren J et al (1995) High-resolution structures of HIV-1 RT from four RT-inhibitor complexes. Nat Struct Biol 2:293–302

26. Xia Q, Radzio J, Anderson KS, Sluis-Cremer N (2007) Probing nonnucleoside inhibitor-induced active-site distortion in HIV-1 reverse transcriptase by transient kinetic analyses. Protein Sci 16:1728–1737
27. Das K et al (1996) Crystal structures of 8-Cl and 9-Cl TIBO complexed with wild-type HIV-1 RT and 8-Cl TIBO complexed with the Tyr181Cys HIV-1 RT drug-resistant mutant. J Mol Biol 264:1085–1100
28. Das K, Martinez SE, Bauman JD, Arnold E (2012) HIV-1 reverse transcriptase complex with DNA and nevirapine reveals non-nucleoside inhibition mechanism. Nat Struct Mol Biol 19:253–259
29. Bec G et al (2013) Thermodynamics of HIV-1 reverse transcriptase in action elucidates the mechanism of action of non-nucleoside inhibitors. J Am Chem Soc 135:9743–9752
30. Das K et al (2007) Crystal structures of clinically relevant Lys103Asn/Tyr181Cys double mutant HIV-1 reverse transcriptase in complexes with ATP and non-nucleoside inhibitor HBY 097. J Mol Biol 365:77–89
31. Das K, Arnold E (2013) HIV-1 reverse transcriptase and antiviral drug resistance. Part 1. Curr Opin Virol 3:111–118
32. Das K, Arnold E (2013) HIV-1 reverse transcriptase and antiviral drug resistance. Part 2. Curr Opin Virol 3:119–128

Chapter 7
Protein-Ligand Interactions as the Basis for Drug Action

Gerhard Klebe

Abstract Lead optimization seeks for conclusive parameters beyond affinity to profile drug-receptor binding. One option is to use thermodynamic signatures since different targets require different mode-of-action mechanisms. Since thermodynamic properties are influenced by multiple factors such as interactions, desolvation, residual mobility, dynamics, or local water structure, careful analysis is essential to define the reference point why a particular signature is given and how it can subsequently be optimized. Relative comparisons of congeneric ligand pairs along with access to structural information allow factorizing a thermodynamic signature into individual contributions.

7.1 Introduction

In a drug development program a lead scaffold, possibly discovered by high-throughput screening [1], virtual computer screening [2] or by a fragment-based lead approach, is optimized from milli via micro to nanomolar binding [3–5]. This optimization is performed by either "growing" the initially discovered scaffold into a binding site, or by exchanging functional groups at its basic skeleton by other, purposefully selected bioisosteric groups. These modifications are intended to increase the binding affinity of the small-molecule ligand toward the target protein and they usually result in an increase of the molecular mass of the candidate molecules to be improved.

G. Klebe (✉)
Department of Pharmaceutical Chemistry, University of Marburg, Marbacher Weg 6, D35032 Marburg, Germany
e-mail: klebe@mailer.uni-Marburg.de

© Springer Science+Business Media Dordrecht 2015   83
G. Scapin et al. (eds.), *Multifaceted Roles of Crystallography in Modern Drug Discovery*, NATO Science for Peace and Security Series A: Chemistry and Biology, DOI 10.1007/978-94-017-9719-1_7

7.2 How to Measure and Rank "Affinity"

To quantify this optimization process, the binding of a ligand to its target protein is measured [6]. Usually the so-called binding constant is determined under the conditions of a chemical equilibrium, which is literally taken, either the dissociation constant K_d or its inverse, the association constant K_a. They indicate what portion of a ligand is bound to the protein according to the underlying law-of-mass. With enzymes usually the so-called inhibition constant K_i is determined in a kinetic enzyme assay. The turn-over of an appropriate substrate is followed concentration dependent. At low substrate concentration, it determines the dependence of the inhibitory concentration on the change in the reaction rate of the enzymatic turnover. Although K_i is not exactly defined as a dissociation constant, K_i, K_d, and K_a are usually referred to interchangeably and represent a kind of strength of the interaction between protein and ligand.

Frequently, instead of the binding constant a so-called IC_{50} value is recorded. This value is characterized by the ligand concentration at which the protein activity has decreased to half of the initial amount. In contrast to the K_i value, the IC_{50} value depends on the concentrations of the enzyme and the substrate used in the enzyme reaction. The obtained value is affected by the affinity of the substrate for the enzyme, as substrate and inhibitor compete for the same binding site. Using the Cheng-Prusoff equation IC_{50} values can be transformed into binding constants [7].

7.3 Affinity: A Thermodynamic Equilibrium Entity Composed by Enthalpy and Entropy

The binding constant can be logarithmically related under constant pressure and standard conditions to thermodynamic properties such as the Gibbs binding free energy ΔG, which itself partitions into an enthalpic and entropic binding contribution, whereby the latter is weighted by the absolute temperature at which the recorded process is determined [8, 9]. The enthalpy reflects the energetic changes during complex formation and can be linked to the interactions associated with the various steps important for the generation of the protein-ligand complex [8]. However, the changes in enthalpy are not the entire answer as to why such a complex is actually formed. In addition, it is important to consider changes in the ordering parameters. This involves how a particular amount of energy is distributed over the multiple degrees of freedom of a given molecular system. This comprises the ligand and the protein prior to complex formation, the formed protein-ligand complex and, important enough, all changes that occur with water and the various components solvated in the water environment (such as buffer compounds or ions to balance the charge inventory in the local environment). Only if this entire system transforms on the whole into a less-ordered state, which corresponds to a situation of increased entropy, a particular process such as the formation of a protein-ligand complex will spontaneously occur. Important enough the entropic component is weighted with temperature. It matters a great deal whether the entropy of a system is changed at low temperature, where all particles are largely in an ordered state, or whether it occurs at high temperature where the disorder is already significantly enhanced. Spontaneously occurring processes are characterized by a negative value for ΔG. Energetically favorable, exothermic processes are defined by a negative enthalpy contribution. If entropy increases, a positive contribution is recorded; however, because the entropic term $T\Delta S$ is considered with a negative sign, an increase in the entropy will cause a decrease in the Gibbs free energy and therefore an increase in binding affinity. A detailed discussion of the various interactions possible to be formed between a protein and a ligand can be found in Ref. [8].

7.4   If a Complex Forms: Two Particles Merge into One

Prior to complex formation, protein and ligand are separately solvated and move freely in the bulk solvent phase. Upon complex formation the two independent particles merge into one species. By this they sacrifice their independent rotational and translational degrees of freedom as two independent particles reduce to one [10]. This loss of about 15–20 kJ/mol is associated with a price in Gibbs free energy to be afforded. This value has been nicely confirmed by a study of Nazare et al. who studied binding of two non-overlapping fragments to FXa [11] and by Borsi et al. [12] who investigated the assembly of an acethydroxamate and a benzenesulfonamide fragment as a potent MMP-12 inhibitor. Comparing the binding affinity of the two individual fragments with that of the merged supermolecule reveals a value of approximately 14–15 kJ/mol. These values match very well with the price to be paid for the loss of degrees of freedom for merging two into one particle.

7.5   How Gibbs Free Energy Factorizes into Enthalpy and Entropy

This fact also sets a lower affinity limit to be expected for complex formation. Only if the newly assembled complex experiences interactions, which will overcome this intrinsic lower barrier of about 15 kJ/mol, a complex can be observed. This finding is nicely reflected by a compilation published by Olsson et al. [13]. The authors have collected the available thermodynamic data in literature and mapped the information in a ΔH versus $-T\Delta S$ diagram (Fig. 7.1). The main diagonal in the $\Delta H/-T\Delta S$ plot corresponds to the observed data scatter in the Gibbs free energy, which covers a range from approx. -15 to -60 kJ/mol. This distribution reflects the range accessible for ligand optimization from milli- to subnano-molar affinity. The diagonal perpendicular to the ΔG distribution reflects the mutual scatter of enthalpy

Fig. 7.1 Thermodynamic data of protein-ligand complexes measured by ITC and plotted in a ΔH versus $-T\Delta S$ diagram. The change in Gibbs free energy is shown along the main diagonal (*dotted line*), perpendicular the scatter in enthalpy and entropy is indicated. In the *dark grey* area enthalpic binding, in the *light gray* area entropic binding preveals. Ligands from medicinal chemistry programs (Δ) tend towards entropically driven binding with increasing affinity (*lower right*) (The figure was adapted from Ref. [13])

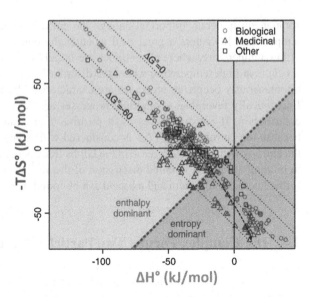

and entropy with opposing contributions to ΔG. As this distribution spreads over a very large range, it discloses an intrinsic enthalpy/entropy compensation that must be in operation, leading to the rather small scatter in ΔG.

The space covered in the enthalpy/entropy diagram can be split into an area where enthalpic binding contributions prevail (dark gray) and an opposing one where entropic contributions (light gray) dominate. It is remarkable to note that ligands originating from medicinal chemistry optimization tend toward enhanced entropic binding profile with growing potency. This immediately calls for the question whether a more enthalpically or entropically driven binding is desired [14–19] and whether such a binding profile of a ligand to be developed can be designed at will [20]? The immanent enthalpy/entropy compensation already suggests that both properties are interdependent, but can they be optimized independently? Most efficient ΔG optimization could be achieved if ΔH and $-T\Delta S$ could be enhanced simultaneously; however, is such a strategy achievable without getting stuck in an enthalpy/entropy compensation trap? Even though there is no physical law, which argues for mutual enthalpy/entropy compensation many considerations on the molecular level suggest that the two opponents will at least partially cancel out [21]. However, strong enthalpic interactions will fix a ligand at the binding site, which is entropically unfavorable. In contrast, pronounced residual mobility in the bound state is entropically beneficial, as a smaller amount of degrees of freedom is lost upon complex formation. Nonetheless, the quality of the formed interactions will be less efficient leading to a minor enthalpic contribution.

7.6   What Profile Is Required: Enthalpy Versus Entropy Driven Binding

This suggests that it is obviously difficult to optimize both properties independently and the tailored design of a predominantly enthalpic or entropic binder represents a major challenge. Notwithstanding, different targets require ligands with different thermodynamic profiles. A CNS drug needs different properties compared to a drug addressing an extracellular target, e.g. in the blood stream. High target selectivity can be of utmost importance, to avoid undesirable side effects, in contrast, promiscuous binding to several members of a protein family can be essential to completely down-regulate a particular biochemical pathway, e.g. in case of kinases, or to achieve a well-balanced binding profile at a given GPCR. In case of viral or bacterial targets, rapid mutational changes can create resistance against a potent ligand. The strategies followed by the pathogens span from steric mismatch in the active site to changes in the protein dynamics to diminish affinity of a bound active agent [22, 23]. As the molecular foundations of these mechanisms are quite distinct well-tailored thermodynamic signatures are required to escape resistance. Freire et al. have suggested improved susceptibility to resistance mutations for ligands optimized enthalpically as they still exhibits sufficient flexibility to evade geometrical modifications of the target protein upon mutational variations [19, 24]. However, equally well ligands binding with entropic advantage due to high residual mobility allowing for multiple binding modes might provide some benefit to escape resistance development. This has remarkably been demonstrated by the superior resistance susceptibility of dapivirine or etravirine over other compounds inhibiting HIV reverse transcriptase [25]. The two inhibitors are characterized by the ability to reorient into alternative binding modes. A firm mapping of the optimal thermodynamic profile to the requirement of a given target is yet not evident and subject to current research.

Drug development based on rational concepts requires detailed understanding of the interactions of a small molecule drug with its target protein. Therefore, increasingly structural and thermodynamic properties of ligand-protein binding in terms of enthalpy/entropy profiles are correlated [26]. It has been proposed to use such profiles to support the decision making process which ligands to take as lead candidates to the next level of development [14–20]. From a theoretical point of view it appears promising and advisable to focus on the most enthalpic binders, as optimization steps governed by entropic factors will be followed unavoidably during late stage optimization. However, at this stage the reasons for a resulting thermodynamic binding signature must be fully characterized to correctly assign 'largest enthalpic efficiency' to a prospective lead.

7.7 Isothermal Titration Calorimetry: Access to Thermodynamic Data

The method of choice to record thermodynamic data is isothermal titration calorimetry (ITC). It provides direct access to ΔG and ΔH in one single experiment, $T\Delta S$ is calculated from their numerical difference. Any error or deficiency in the measurement of these two properties will cause an inevitable $\Delta H/T\Delta S$ compensation, apart from the heavily discussed intrinsic enthalpy/entropy compensation in biological systems (s. above). Not to get trapped in an error-prone compensation, thorough analysis and correction of superimposed effects of ITC data has to be performed and it is highly advisable to only correlate matching ligand pair series relative to each other.

7.8 Contributions to the Thermodynamic Profile: H-bonds and Lipophilic Contacts

Hydrogen bonding usually relates to an enthalpic signal which increases once growing charges of the interacting functional groups are involved [27–29]. However, with larger charges also a detrimental entropic contribution is experienced which reduces, due to enthalpy/entropy compensation, the overall free energy contribution of an H-bond. Lipophilic contacts buried upon complex formation result in an increasing entropic signal, but only, if ordered water molecules are displaced from the binding pocket [29–31]. Mobile water molecules displaced upon ligand binding can also give rise to a more enthalpy-driven binding [32, 33]. If no permanent and strong charges of the interacting species are involved, the release or pick-up of water molecules upon ligand binding seems to be virtually balanced out in the Gibbs free energy inventory, but huge effects are experienced with respect to the enthalpy/entropy partitioning [31, 34]. This observation demonstrates that the sole determination of free energy will hardly unravel involvement of water molecules in binding. This also explains why surprisingly many computer modeling approaches can still generate reasonable ΔG predictions neglecting water, but geometries will be predicted incorrectly.

7.9 Preorganization and Rigidization of Ligands, Cooperative Effects

Ligand pre-organization and rigidization of the protein-bound conformation can result in large beneficial free energy contributions, mainly due to an entropic advantage. These generalized signatures often become only transparent once a congeneric series of ligands is evaluated as the overall thermodynamic profile of the binding process can be superimposed by multiple effects arising from changes in the dynamics of either protein and/or ligand, rearrangements of the protein and most important by changes of the solvation pattern of discrete water molecules. Furthermore, puzzling cooperativity between hydrogen bonding and hydrophobic contacts can be given resulting from changes in the dynamics of protein-ligand complexes and modulations of residual solvation pattern [35–37].

7.10   The Role of Water in Ligand Binding and Thermodynamics

Remarkable effects arise from rearrangements of surface water molecules wrapping around newly formed protein-ligand complexes [37–39]. Water networks span across the newly created complex surfaces and exhibit geometric and energetic fits of deviating quality. Ideal fit results in an affinity enhancement of the bound ligand; imperfect and fragmented water networks reduce affinity of the bound ligand. Moreover, such changes are reflected by major modulations of the enthalpy/entropy signature and easily provoke a mutual ΔH vs. $T\Delta S$ shift of ± 5–10 kJ/mol. If the residual solvation pattern takes such an enormous impact on the thermodynamic signature, classification of a given ligand as "more enthalpic" or "more entropic" binder appears rather meaningless without full information about the structural properties of the formed complex e.g. by means of high-resolution crystal structure analysis. Only then the thermodynamic profile can support the decision making process which ligand to take to the next level of development. Nonetheless, deviating thermodynamic profiles recorded across congeneric ligand series unambiguously indicate differences in the binding patterns, be it for deviations in binding poses, residual solvation patterns or intrinsic dynamics.

ITC measurements can also help to record whether a change in protonation state occurs when a ligand binds to a protein. Therefore the thermodynamic parameters have to be measured from different buffer conditions. The obtained results can be used to drive the tailored design of pK_a properties of ligands [40]. If in a congeneric ligand series thermodynamic data show an unexpected shift between enthalpy and entropy even though the Gibbs free energy of binding remains virtually unchanged among the different ligands, usually a remarkable effect or change of the system is superimposed to the binding event. Clearly such effects cannot be seen considering solely affinity data. Since the involvement of water molecules in the binding interface takes mostly minor impact of the free energy but huge effects are seen in the enthalpy/entropy inventory, thermodynamic data can uncover the importance of water on ligand binding. For the same reasons the influence of water often foils a straight forward comparison of thermodynamic signatures across ligand series without having access to structural information in parallel, as the entrapping or release of a single water molecule can easily invert the thermodynamic profile. Through thermodynamic data impressive cooperative effects resulting either from deviating dynamic behaviour of the formed complexes [35, 36] or changes in the surface water structure became evident [37–39]. These effects became only obvious by carefully analyzing the deviating trends in the thermodynamic profiles of the formed complexes. Finally, the partitioning of the Gibbs free energy of binding in enthalpy and entropy can help to understand flat structure-activity relationships and distinguish ligand binding to deviating conformations of the target protein, an effect not to be unravelled purely considering affinity data [41].

Even though we can establish some general rules how to fight enthalpy/entropy compensation and substantiate the reasoning why to start with leads of "high enthalpic efficiency", the overall binding event shows many additional phenomena giving rise to an undesired compensation. It remains in question whether they can always be fully elucidated and avoided. But they provide an explanation why it is still not trivial and straight forward possible to factorize a thermodynamic signature into individual contributions that can be attributed to single interactions formed between a lead candidate and its target protein.

References

1. Mayr LM, Bojanic D (2009) Novel trends in high-throughput screening. Curr Opin Pharmacol 9:580–588
2. Klebe G (2006) Virtual ligand screening: strategies, perspectives, and limitations. Drug Discov Today 11:580–594
3. Wermuth CG (2003) Chapter 18: application of strategies for primary structure-activity relationship exploration. In: Wermuth CG (ed) The practice of medicinal chemistry. Elsevier, Amsterdam
4. Blundell TL, Jhoti H, Abell C (2002) High-throughput crystallography for lead discovery in drug design. Nat Rev Drug Discov 2:45–53
5. Kloe GE, de Bailey D, Leurs R, Esch IJP (2009) Transforming fragments into candidates: small becomes big in medicinal chemistry. Drug Discov Today 14:630–646
6. Ajay, Murcko MA (1995) Computational methods to predict binding free energy in ligand-receptor complexes. J Med Chem 38:4953–4967
7. Cheng YC, Prusoff WH (1973) Relationship between the inhibition constant (K_i) and the concentration of inhibitor which causes 50 per cent inhibition (I_{50}) of an enzymatic reaction. Biochem Pharmacol 22:3099–3108
8. Klebe G (2013) Drug design, Chapter 4, Springer Reference, Heidelberg, New York, Dordrecht, London
9. Chaires JB (2008) Calorimetry and thermodynamics in drug design. Annu Rev Biophys 37:135–151
10. Murray CW, Verdonk ML (2002) The consequences of translational and rotational entropy lost by small molecules on binding to proteins. J Comput Aided Mol Des 16:741–753
11. Nazare M, Matter H, Will DW, Wagner M, Urmann M, Czech J, Schreuder H, Bauer A, Ritter K, Wehner V (2012) Fragment deconstruction of small, potent factor Xa inhibitors: exploring the superadditivity energetic of fragment linking in protein-ligand complexes. Angew Chem Int Ed 51:905–911
12. Borsi V, Calderone V, Fragai M, Luchinat C, Sarti N (2010) Entropic contribution to the linking coefficient in fragment-based drug design: a case study. J Med Chem 53:4285–4289
13. Olsson TSG, Williams MA, Pitt WR, Ladbury JE (2008) The thermodynamics of protein-ligand interactions and solvation: insights for ligand design. J Mol Biol 384:1002–1017

14. Ladbury JE, Klebe G, Freire E (2010) Adding calorimetric data to decision making in lead discovery: a hot tip. Nat Rev Drug Discov 9:23–27
15. Hann MM, Kerserü GM (2011) Finding the sweet spot: the role of nature and nurture in medicinal chemistry. Nat Rev Drug Discov 11:355–365
16. Ferenczy GG, Kerserü GM (2010) Thermodynamics guided lead discovery and optimization. Drug Discov Today 15:919–932
17. Reynolds CH, Holloway MK (2011) Thermodynamics of ligand binding and efficiency. ACS Med Chem Lett 2:433–437
18. Ferenczy GG, Keserü GM (2012) Thermodynamics of fragment binding. J Chem Inf Model 52:1039–1045
19. Freire E (2008) Do enthalpy and entropy distinguish first in class from best in class? Drug Discov Today 13:869–874
20. Freire E (2009) A thermodynamic approach to the affinity optimization of drug candidates. Chem Biol Drug Des 74:468–472
21. Dunitz JD (2003) Win some, lose some: enthalpy-entropy compensation in weak intermolecular interactions. Chem Biol 2:709–712
22. Weber IT, Agniswamy J (2009) HIV-1 protease: structural perspective on drug resistance. Viruses 1:1110–1136
23. Ali A, Bandaranayake RM, Cai Y, King NM, Kolli M, Mittal S, Murzycki JF, Nalam MNL, Nalivaika EA, Özen A, Prabu-Jeyabalan MM, Thayer K, Schiffer CA (2010) Molecular basis for drug resistance in HIV-1 protease. Viruses 2:2509–2535
24. Ohtaka H, Freire E (2005) Adaptive inhibitors of the HIV-1 protease. Prog Biophys Mol Biol 88:193–208
25. Das K, Lewi PJ, Hughes SH, Arnold E (2005) Crystallography and the design of anti-AIDS drugs: conformational flexibility and positional adaptability are important in the design of non-nucleoside HIV-1 reverse transcriptase inhibitors. Prog Biophys Mol Biol 88:209–231
26. Martin SF, Clements JH (2013) Correlating structure and energetics in protein-ligand interactions: paradigms and paradoxes. Annu Rev Biochem 82:267–293
27. Steuber H, Heine A, Klebe G (2007) Structural and thermodynamic study on aldose reductase: nitro-substituted inhibitors with strong enthalpic binding contribution. J Mol Biol 368:618–638
28. Steuber H, Czodrowski P, Sotriffer CA, Klebe G (2007) Tracing changes in protonation: a prerequisite to factorize thermodynamic data of inhibitor binding to aldose reductase. J Mol Biol 373:1305–1320
29. Baum B, Mohamed M, Zayed M, Gerlach C, Heine A, Hangauer D, Klebe G (2009) More than a simple lipophilic contact: a detailed thermodynamic analysis of non-basic residues in the S1 pocket of thrombin. J Mol Biol 390:56–69
30. Biela A, Khayat M, Tan H, Kong J, Heine A, Hangauer D, Klebe G (2012) Impact of ligand and protein desolvation on ligand binding to the S1 pocket of thrombin. J Mol Biol 418:350–366
31. Biela A, Sielaff F, Terwesten F, Heine A, Steinmetzer T, Klebe G (2012) Ligand binding stepwise disrupts water network in thrombin: enthalpic and entropic changes reveal classical hydrophobic effect. J Med Chem 55:6094–6110
32. Englert L, Biela A, Zayed M, Heine A, Hangauer D, Klebe G (2010) Displacement of disordered water molecules from the hydrophobic pocket creates enthalpic signature: binding of phosphonamidate to the S1'-pocket of thermolysin. Biochim Biophys Acta 1800:1192–1202
33. Homans SW (2007) Water, water everywhere – except where it matters. Drug Discov Today 12:534–539
34. Petrova T, Steuber H, Hazemann I, Cousido-Siah A, Mitschler A, Chung R, Oka M, Klebe G, El-Kabbani O, Joachimiak A, Podjarny A (2005) Factorizing selectivity determinants of inhibitor binding toward aldose and aldehyde reductases: structural and thermodynamic properties of the aldose reductase mutant Leu300Pro-fidarestat complex. J Med Chem 48:5659–5665

35. Baum B, Muley L, Smolinski M, Heine A, Hangauer D, Klebe G (2010) Non-additivity of functional group contributions in protein-ligand binding: a comprehensive study by crystallography and isothermal titration calorimetry. J Mol Biol 397:1042–1057

36. Muley L, Baum B, Smolinski M, Freindorf M, Heine A, Klebe G, Hangauer D (2010) Enhancement of hydrophobic interactions and hydrogen bond strength by cooperativity: synthesis, modeling, and molecular dynamics simulations of a series of thrombin inhibitors. J Med Chem 53:2126–2135

37. Biela A, Betz M, Heine A, Klebe G (2012) Water makes the difference: rearrangement of water solvation layer triggers non-additivity of functional group contributions in protein-ligand binding. ChemMedChem 7:1423–1434

38. Biela A, Nasief NN, Betz M, Heine A, Hangauer D, Klebe G (2013) Dissecting the hydrophobic effect on the molecular level: the role of water, enthalpy, and entropy in ligand binding to thermolysin. Angew Chem Int Ed 52:1822–1828

39. Krimmer S, Betz M, Heine A, Klebe G (2014) Methyl, ethyl, propyl, butyl: futile but not for water, as the correlation of structure and thermodynamic signature shows in a congeneric series of thermolysin inhibitors. ChemMedChem 9:833–846

40. Neeb M, Czodrowski P, Heine A, Barandun LJ, Hohn C, Diederich F, Klebe G (2014) Chasing Protons: How ITC, mutagenesis and pKa calculations trace the locus of charge in ligand binding to a tRNA-binding enzyme. J Med Chem 57:5554–5565

41. Neeb M, Betz M, Heine A, Barandun LJ, Hohn C, Diederich F, Klebe G (2014) Beyond affinity: enthalpy-entropy factorization unravels complexity of a flat structure-activity relationship for inhibition of tRNA-modifying enzyme. J Med Chem 57:5566–5578

Chapter 8
The Protein Data Bank: Overview and Tools for Drug Discovery

Helen M. Berman, Peter W. Rose, Shuchismita Dutta, Christine Zardecki, and Andreas Prlić

Abstract The increasing size and complexity of the three dimensional (3D) structures of biomacromolecules in the Protein Data Bank (PDB) is a reflection of the growth in the field of structural biology. Although the PDB archive was initially used only in the field of structural biology, it has grown to become a valuable resource for understanding biology at a molecular level and is critical for designing new therapeutic options for various diseases. The many uses of the PDB archive depend upon on the tools and resources for both data management and for data access and analysis.

8.1 Introduction

The field of structural biology began in the late 1950s as scientists started to decipher the three dimensional (3D) structures of proteins. Structure determination of myoglobin [1, 2] followed closely by that of hemoglobin [3, 4] earned Perutz and Kendrew Nobel prizes in 1962. Soon members of the scientific community recognized how strong research advances could be made through a shared, public archive of data from these experiments [5, 6]. In 1971, following a meeting at Cold Spring Harbor, the Protein Data Bank (PDB) was established with seven structures [7].

H.M. Berman (✉) • S. Dutta • C. Zardecki
RCSB Protein Data Bank, Department of Chemistry and Chemical Biology and Center
for Integrative Proteomics Research, Rutgers, The State University of New Jersey, Piscataway,
NJ 08854-8087, USA
e-mail: berman@rcsb.rutgers.edu

P.W. Rose • A. Prlić
RCSB Protein Data Bank, San Diego Supercomputer Center, University of California San Diego,
La Jolla, CA 92093-0743, USA

© Springer Science+Business Media Dordrecht 2015
G. Scapin et al. (eds.), *Multifaceted Roles of Crystallography in Modern Drug
Discovery*, NATO Science for Peace and Security Series A: Chemistry and Biology,
DOI 10.1007/978-94-017-9719-1_8

Today, the PDB archive contains more than 100,000 structures and is managed by the Worldwide Protein Data Bank (wwPDB, wwpdb.org), a consortium of groups that host deposition, annotation, and distribution centers for PDB data and collaborate on a variety of projects and outreach efforts [8, 9]. While the PDB data is available as a single archive, wwPDB data centers present unique tools, resources and views of the data to facilitate scientific inquiry and analysis.

8.2   Overview

8.2.1   PDB Data

The primary data archived in the PDB are the 3D atomic coordinates of biological molecules determined using experimental methods such as X-ray crystallography, Nuclear Magnetic Resonance (NMR) and Electron Microscopy (3D EM). In addition to coordinate data PDB also archives several descriptive metadata items such as the primary citation, polymer sequence, chemical information about the ligands and macromolecules, some experimental details, and structural descriptors. Experimental data used to derive these structures (e.g. structure factors, restraints and chemical shifts) are made available, along with 3DEM map data [10].

All information regarding a particular structure is linked to an identifier (PDB ID). The original file format used to represent PDB was established 40 years ago and has very recently been replaced by PDBx/mmCIF. This newer format is computer readable and unlike the older format can accommodate large complex structural data. The PDBx/mmCIF Data Exchange Dictionary [11] consolidates content from a variety of crystallographic data dictionaries and includes extensions describing NMR, 3DEM, and protein production data. Internal data processing, annotation, and database management operations rely on the PDBx/mmCIF dictionary content and corresponding file format. As the PDBx/mmCIF file format is very extensible, it can expand and grow to support new types of information. Recently, the developers of X-ray structure determination packages have adapted PDBx/mmCIF as their standard format.

8.2.2   Data Deposition and Annotation

Once a structure has been determined, it is deposited into the PDB for processing and annotation by the wwPDB. Until recently, multiple different systems for deposition and annotation made data uniformity and exchange difficult. In the new wwPDB Common Deposition & Annotation (D&A) system, launched in 2014, data are easily transferred and shared. In addition, many aspects of the deposition and annotation practices have been improved enabling efficiency and accuracy (Fig. 8.1).

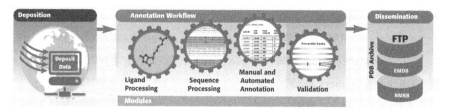

Fig. 8.1 wwPDB Common Deposition and Annotation System for PDB, EMDB and BMRB data. In this pipeline, data are submitted to the PDB using a single interface and then processed and annotated by the wwPDB using a series of focused modules [16, 32]. Data are released into the PDB FTP archive at ftp://ftp.wwpdb.org on a weekly basis

Highly qualified biocurators in the wwPDB data processing centers annotate each PDB entry to ensure accurate representation of both the structure and experiment. They review polymer sequences, small molecule chemistry, cross references to other databases, experimental details, correspondence of coordinates with primary data, protein conformation, biological assemblies, and crystal packing. During the annotation process, the wwPDB biocurators communicate with the entry authors (depositors) to make sure the data are represented in the best way possible.

To help ensure the accuracy of PDB entries, deposited data are compared with community-accepted standards during the process of validation. Method-specific Validation Task Forces (VTF) comprising of experts in X-ray Crystallography [12], NMR [13], 3DEM [14], and Small Angle Scattering [15] were convened by the wwPDB to develop consensus on validation that should be performed, and to identify software applications for validation. The VTF recommendations are now implemented in the wwPDB data processing procedures and suitable tools have been developed as part of the wwPDB Common Deposition & Annotation System.

Depositors are provided with detailed reports that include the results of data consistency, geometric and experimental data validation [16]. These reports, available as PDFs, provide an assessment of structure quality while maintaining the confidentiality of the coordinate data. Graphical depictions allow facile assessments of the overall quality as well as sequence specific features (Fig. 8.2). Currently, these wwPDB validation reports are required by several journals for manuscript review, including *eLife*, *The Journal of Biological Chemistry*, and the journals of the International Union of Crystallography. The wwPDB encourages all journal editors and referees to incorporate these reports in the manuscript submission and review process.

8.2.3   Data Distribution

The PDB archive (ftp://ftp.wwpdb.org) is updated weekly. Contents of the ftp site include experimentally determined coordinate data files, related experimental data (structure factors, constraints, and chemical shifts) and 3DEM map data. The

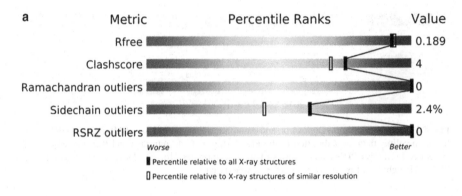

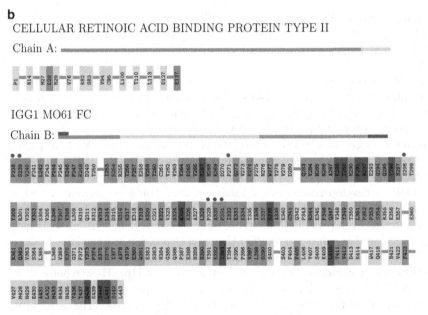

Fig. 8.2 Graphics included in the Validation Reports produced by the wwPDB. These reports, made available as PDFs, provide an assessment of structure quality while maintaining the confidentiality of the coordinate data. (**a**) The "slider" graphic gives an indication of the quality of the determined structure as compared with previously deposited PDB entries using several important global quality indicators. (**b**) Residue-property plots indicate quality information for proteins and nucleic acids on a per-residue basis. Two images are displayed for each molecule. In the *top image*, the *green*, *yellow*, *orange* and *red* segments indicate the fraction of residues with 0, 1, 2 and 3 or more types of model-only quality criteria with outliers. In the *bottom image*, the *red circle* (if present) indicates the fraction of residues that have an unusual fit to the density (RSRZ outliers) (Color figure online)

ftp site also contains the data dictionaries and external reference files (ERFs) used to describe PDB data, including the PDBx/mmCIF dictionary, the Chemical Component Dictionary (CCD) that contains detailed chemical descriptions for standard and modified amino acids/nucleotides, small molecule ligands, and solvent molecules, and the Biologically Interesting Molecule Reference Dictionary (BIRD) that contains information about biologically interesting peptide-like antibiotic and inhibitor molecules in the PDB archive [17].

Each wwPDB member organization maintains websites with different views of the data and different services. These websites are RCSB PDB (US) at rcsb.org [18], Protein Data Bank in Europe (PDBe, United Kingdom) at pdbe.org [19], Protein Data Bank Japan (PDBj) at pdbj.org [20], and the BioMagResBank (BMRB, US) at bmrb.wisc.edu [21].

8.2.4 Growth of the PDB Archive

The number of structures contained in the archive has grown over the past ~40 years since the creation of the PDB. In addition to structures determined by X-ray crystallography, the archive includes structures determined using NMR spectroscopy and 3D electron microscopy (3D EM) (Fig. 8.3 a–c). It is worth noting that the growth in the number of cryoEM maps is an indicator of the expected high growth rate of cryoEM-derived models that are being deposited into the PDB.

In addition, the complexity of structures deposited has increased as evidenced by growth in the number of polymers chains within each structure and the molecular weight (Fig. 8.4a, b). By reviewing the content within the PDB it is possible to see the evolution of types of methods used to determine structures. Whereas in the 1970s only relatively small structures could be studied, now we have many examples of macromolecular machines [22, 23] (Fig. 8.5). Most recently, structures have been determined using several different methods. These hybrid models are the subject of much discussion as to how to best evaluate and archive them.

8.3 RCSB PDB Resources for Drug Discovery

In addition to biological macromolecules (proteins and nucleic acids), ~73 % of PDB entries include one or more ligands. Some of these ligands are simple, such as ions, cofactors, inhibitors, and drugs [22]. More than 1,000 PDB structures contain peptide-like inhibitors and antibiotics [17]. These ligand-bound complexes highlight the overall shapes and key functional regions of the relevant biological molecules and lay the foundations for designing molecules that can alter the function. In the 1980s when Acquired Immunodeficiency Syndrome (AIDS) was rapidly spreading through the world, structural studies of Human Immunodeficiency Virus (HIV) proteins were critical in designing specific inhibitors that have led to the development of clinically important drugs for treating HIV infection [24–26]. Similarly, there have been many studies of antibiotics that target the ribosome [27, 28].

a

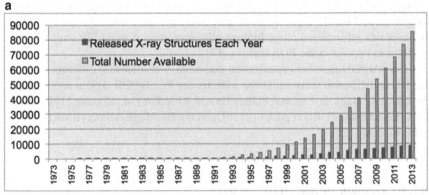

b

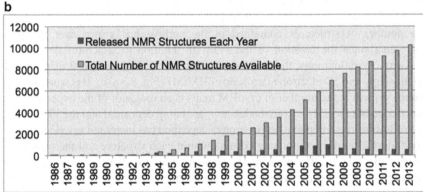

c

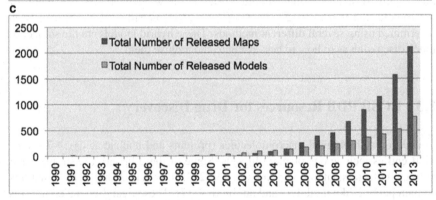

Fig. 8.3 Growth of the number of structures available in the PDB archive by experimental method: (**a**) X-ray crystallography, (**b**) NMR, (**c**) 3DEM

While biological polymers (proteins and nucleic acids) can be queried in the PDB by protein or gene name or its sequence, the RCSB Protein Data Bank website provides a number of resources that facilitate drug discovery-related research [29]. The following sections provide a brief description of these tools.

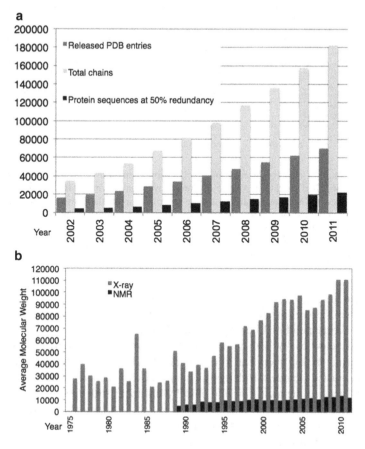

Fig. 8.4 Growth of the size and complexity of the structures available in the PDB archive. (**a**) the number of PDB entries, total related polymer chains, and protein sequences (with 50 % redundancy as calculated using blastclust [33]) available in the archive each year; (**b**) Average molecular weight of entries released each year for structures determined by X-ray crystallography (for the asymmetric unit; in *grey*) and NMR (in *black*). Calculations excluded water and counted extremely large structures as single entries. For viruses and entries that used non-crystallographic symmetry (NCS), molecular weights for the full asymmetric unit were calculated by multiplying the molecular weight of the explicit polymer chains by the number of NCS operators. The large increase shown in 1984 was due to the release of the tomato bushy stunt virus 2tbv [34] (Figures reprinted from [22])

8.3.1 Ligand Search

The most common uses of the RCSB PDB website are simple searches using the top search box on the RCSB PDB website. An autocomplete feature is available that can help guide the user to specific matches in the archive and provide relevant results. After typing a few letters in the top search bar, a suggestion box opens and organizes result sets in different categories. Each suggestion includes the number of

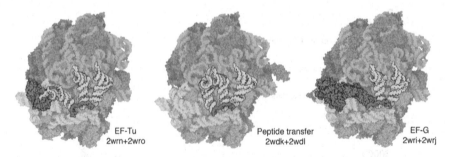

Fig. 8.5 Example of a macromolecular machine: ribosome complexes (PDB IDs 2wrn, 2wro [35], 2wdk, 2wdl [36], 2wri, 2wrj [37]. Atomic structures have been determined for ribosomes engaged in most aspects of mRNA translation. These three structures capture the ribosome in distinct phases of elongation: *left*, binding of a new tRNA assisted by elongation factor Tu; *middle*, the peptide transfer reaction; and *right*, stepping to the next reading frame by binding of elongation factor G. Image from the RCSB PDB *Molecule of the Month* feature on the Ribosome (doi: 10.2210/rcsb_pdb/mom_2010_1) and reprinted from [23]

results and links to the set of matching structures. For example, by entering the drug brand name "Glivec" or the generic name "Imatinib" the autosuggestion provides a link to the corresponding Ligand Summary Page described below.

Ligand searches by ID, name, synonym, formula, and SMILES string are possible using the top query bar. These queries are also available from the Advanced Search menu and include searching by Chemical Component identifier of the ligand, SMILES strings, chemical formula, and by chemical structure (including exact, substructure, superstructure, and similarity searches). Detailed information about ligands and drug molecules bound to macromolecules are available from the **Ligand Summary** and **Structure Summary** pages.

8.3.2 Ligand Summary Page

Information about the chemistry and structure of all small molecule components found in the PDB is contained in the Chemical Component Dictionary (CCD). The Ligand Summary pages present a report from the CCD are organized into widgets or boxes highlighting different types of hyperlinked information (Fig. 8.6). These widgets provide an overview of the ligand, with links to PDB entries where the component appears as a non-polymer or as a non-standard component of a polymer, links to ligand summary pages for similar ligands and stereoisomers, 2D and 3D visualization, and links to many external resources. Original data provided by the RCSB PDB are listed in blue widgets, whereas data from third parties are displayed in orange widgets.

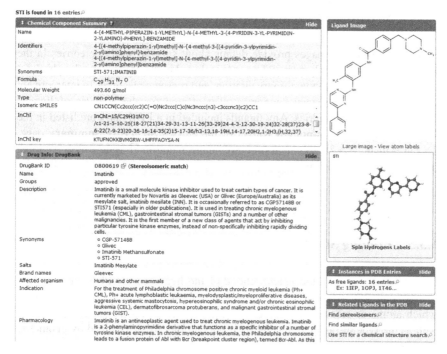

Fig. 8.6 Ligand Summary Page (*top section*) for Imatinib (Glivec). RCSB PDB's Ligand Summary Pages provide information for all of the entries found in the wwPDB's Chemical Component Dictionary. Similar to Structure Summary pages for PDB entries, Ligand Summary Pages are organized into widgets that highlight different types of information, including a Chemical Component Summary that includes name, identifiers, synonyms, and SMILES and InChI information; links to related PDB structures where the ligand appears as a free ligand; links to other Summary Pages for similar ligands and stereoisomers, and links to information about the chemical component at external resources. These summaries can be accessed by performing a ligand search, selecting a ligand from a PDB entry's Structure Summary page, and from the Ligand Hits tab for query results. In the example shown, Glivec is present in 16 PDB entries as a co-crystal structure. Drug annotation is provided by DrugBank [31]

8.3.3   Ligand Summary Reports

For queries that return a set of ligands, the results can be saved as **Ligand Summary Reports** in form of a comma separated value (CSV) file or an Excel spreadsheet. These reports include information about the ligands, such as formula, molecular weight, name, SMILES string, and lists of PDB entries that include the ligand. The report can be expanded to show a sub-table of all PDB entries that contain the ligand as a free ligand and those that contain the ligand as part of a polymer.

8.3.4   Structure Summary Page

Structure Summary pages provide details about specific structure entries in the PDB. It describes all polymers and ligands included in the entry, some details about the experiment, links to the primary citation, and presents resources to interactively visualize the entry. Special support is also offered for the analysis of ligands associated with PDB entries. Any ligands included in a PDB entry are listed in the **Ligand Chemical Component** widget of the entry's **Structure Summary** page. This area displays a 2D chemical structure image, name and formula of each ligand, link to the **Ligand Summary** page, and provides access to 2D and 3D binding site visualization.

8.3.5   Binding Site Visualization

In order to understand the neighborhood of the ligand in the PDB entry and its interactions, 2D interaction diagrams are generated by PoseView [30] and show which atoms or areas of the ligand and the polymer interact with each other, as well as the type of interaction (Fig. 8.7). Interactions are determined by geometric criteria.

Ligand Explorer is a 3D viewer that visualizes the interactions of bound ligands in protein and nucleic acids structures (Fig. 8.8). It has options to turn on the display of interactions including hydrogen bonds, hydrophobic contacts, water mediated hydrogen bonds, and metal interactions. Several types of binding site surfaces can be generated including opaque and transparent solid surfaces, meshes, and dotted surfaces, color coded by hydrophobicity or chain identifier.

8.3.6   Drug and Drug Target Mapping

A detailed mapping of drugs by chemical structure and drug targets by protein sequence is available from the **Drug and Drug Target Mapping** page, which is accessible from the **Search** menu on the RCSB PDB website. Two tables provide access to information about drugs and drug targets from DrugBank [31] that are mapped to PDB entries with each weekly update.

- Drugs Bound to Primary Targets: Lists drugs bound to primary target(s), or a homolog of primary target(s), i.e., co-crystal structures of drugs.
- Primary Drug Targets: Lists primary drug targets in the PDB, regardless if the drug molecule is part of the PDB entry (e.g., apo forms of drug targets, drug target with different bound ligands). Biotherapeutics, such as complexes with monoclonal antibodies, are included.

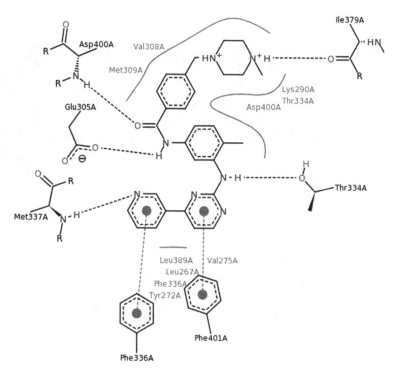

Fig. 8.7 2D macromolecule-ligand interaction diagram of Imatinib (Glivec) bound to Proto-oncogene tyrosine-protein kinase ABL1 (PDB Id: 1OPJ, [38]) generated by PoseView (*black dashed lines*: hydrogen bonds and salt bridges, *green solid lines*: hydrophobic interactions, *green dashed lines*: Pi-Pi interactions) (Color figure online)

These tables can be searched, filtered, sorted, and downloaded as Excel Spread-sheets.

8.4   Summary

The PDB was established in 1971 to archive the experimentally determined 3D structures of biological macromolecules. Today, the archive contains the atomic coordinates and experimental data for more than 100,000 proteins, nucleic acids, and large macromolecular machines. Under the management of the wwPDB collaboration, a new data deposition and annotation tool has been developed to efficiently receive and carefully annotate PDB depositions before public release in the archive.

The RCSB PDB website offers a number of different resources to search, visualize, compare and analyze PDB data. Many of these tools are focused on the study of drug complexes available in the archive.

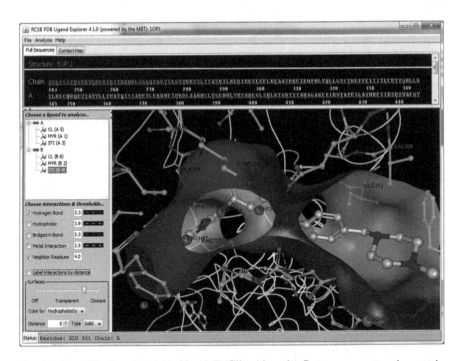

Fig. 8.8 Ligand Explorer 3D view of Imatinib (Glivec) bound to Proto-oncogene tyrosine-protein kinase ABL1 (PDB ID: 1OPJ [38]). The binding pocket is delineated by a surface color-coded by hydrophobicity of the binding site residues (*yellow*: hydrophobic, *blue*: hydrophilic). The vertical cross-section looking into the binding site is transparent and shows the residues linking the drug-binding pocket (Color figure online)

Acknowledgement The RCSB PDB is supported by funds from the National Science Foundation (NSF DBI 1338415), National Institutes of Health, and the Department of Energy (DOE). RCSB PDB is a member of the Worldwide Protein Data Bank.

References

1. Kendrew JC, Bodo G, Dintzis HM, Parrish RG, Wyckoff H, Phillips DC (1958) A three-dimensional model of the myoglobin molecule obtained by x-ray analysis. Nature 181: 662–666
2. Kendrew JC, Dickerson RE, Strandberg BE, Hart RG, Davies DR, Phillips DC, Shore VC (1960) Structure of myoglobin: a three-dimensional Fourier synthesis at 2 A. resolution. Nature 185(4711):422–427
3. Perutz MF, Rossmann MG, Cullis AF, Muirhead H, Will G, North ACT (1960) Structure of haemoglobin: a three-dimensional Fourier synthesis at 5.5 Å resolution, obtained by X-ray analysis. Nature 185:416–422
4. Bolton W, Perutz MF (1970) Three dimensional fourier synthesis of horse deoxyhaemoglobin at 2.8 Ångstrom units resolution. Nature 228(271):551–552

5. Berman HM, Kleywegt GJ, Nakamura H, Markley JL (2013) How community has shaped the Protein Data Bank. Structure 21(9):1485–1491
6. Berman H (2008) The Protein Data Bank: a historical perspective. Acta Crystallogr A: Found Crystallogr 64:88–95
7. Protein Data Bank (1971) Protein Data Bank. Nat New Biol 233:223
8. Berman HM, Henrick K, Nakamura H (2003) Announcing the worldwide Protein Data Bank. Nat Struct Biol 10(12):980
9. Berman HM, Henrick K, Kleywegt G, Nakamura H, Markley J (2012) The worldwide Protein Data Bank. In: Arnold E, Himmel DM, Rossmann MG (eds) International tables for X-ray crystallography, vol F, Crystallography of biological macromolecules. Springer, Dordrecht, pp 827–832
10. Lawson CL, Baker ML, Best C, Bi C, Dougherty M, Feng P, van Ginkel G, Devkota B, Lagerstedt I, Ludtke SJ, Newman RH, Oldfield TJ, Rees I, Sahni G, Sala R, Velankar S, Warren J, Westbrook JD, Henrick K, Kleywegt GJ, Berman HM, Chiu W (2011) EMDataBank.org: unified data resource for CryoEM. Nucleic Acids Res 39:D456–D464
11. Westbrook J, Henrick K, Ulrich EL, Berman HM (2005) 3.6.2 The Protein Data Bank exchange data dictionary. In: Hall SR, McMahon B (eds) International tables for crystallography, vol G. Definition and exchange of crystallographic data. Springer, Dordrecht, pp 195–198
12. Read RJ, Adams PD, Arendall WB III, Brunger AT, Emsley P, Joosten RP, Kleywegt GJ, Krissinel EB, Lutteke T, Otwinowski Z, Perrakis A, Richardson JS, Sheffler WH, Smith JL, Tickle IJ, Vriend G, Zwart PH (2011) A new generation of crystallographic validation tools for the Protein Data Bank. Structure 19(10):1395–1412
13. Montelione GT, Nilges M, Bax A, Güntert P, Herrmann T, Markley JL, Richardson J, Schwieters C, Vuister GW, Vranken W, Wishart D (2013) Recommendations of the wwPDB NMR structure validation task force. Structure 21:1563–1570
14. Henderson R, Sali A, Baker ML, Carragher B, Devkota B, Downing KH, Egelman EH, Feng Z, Frank J, Grigorieff N, Jiang W, Ludtke SJ, Medalia O, Penczek PA, Rosenthal PB, Rossmann MG, Schmid MF, Schroder GF, Steven AC, Stokes DL, Westbrook JD, Wriggers W, Yang H, Young J, Berman HM, Chiu W, Kleywegt GJ, Lawson CL (2012) Outcome of the first electron microscopy validation task force meeting. Structure 20(2):205–214
15. Trewhella J, Hendrickson WA, Sato M, Schwede T, Svergun D, Tainer JA, Westbrook J, Kleywegt GJ, Berman HM (2013) Meeting report of the wwPDB small-angle scattering task force: data requirements for biomolecular modeling and the PDB. Structure 21:875–881
16. Gore S, Velankar S, Kleywegt GJ (2012) Implementing an X-ray validation pipeline for the Protein Data Bank. Acta Crystallogr D68:478–483
17. Dutta S, Dimitropoulos D, Feng Z, Periskova I, Sen S, Shao C, Westbrook J, Young J, Zhuravleva M, Kleywegt G, Berman H (2013) Improving the representation of peptide-like inhibitor and antibiotic molecules in the Protein Data Bank. Biopolymers 101(6):659–668
18. Berman HM, Westbrook JD, Feng Z, Gilliland G, Bhat TN, Weissig H, Shindyalov IN, Bourne PE (2000) The Protein Data Bank. Nucleic Acids Res 28:235–242
19. Gutmanas A, Alhroub Y, Battle GM, Berrisford JM, Bochet E, Conroy MJ, Dana JM, Fernandez Montecelo MA, van Ginkel G, Gore SP, Haslam P, Hatherley R, Hendrickx PM, Hirshberg M, Lagerstedt I, Mir S, Mukhopadhyay A, Oldfield TJ, Patwardhan A, Rinaldi L, Sahni G, Sanz-Garcia E, Sen S, Slowley RA, Velankar S, Wainwright ME, Kleywegt GJ (2014) PDBe: Protein Data Bank in Europe. Nucleic Acids Res 42(1):D285–D291
20. Kinjo AR, Suzuki H, Yamashita R, Ikegawa Y, Kudou T, Igarashi R, Kengaku Y, Cho H, Standley DM, Nakagawa A, Nakamura H (2012) Protein Data Bank Japan (PDBj): maintaining a structural data archive and resource description framework format. Nucleic Acids Res 40:D453–D460
21. Ulrich EL, Akutsu H, Doreleijers JF, Harano Y, Ioannidis YE, Lin J, Livny M, Mading S, Maziuk D, Miller Z, Nakatani E, Schulte CF, Tolmie DE, Kent Wenger R, Yao H, Markley JL (2008) BioMagResBank. Nucleic Acids Res 36:D402–D408

22. Berman HM, Coimbatore Narayanan B, Costanzo LD, Dutta S, Ghosh S, Hudson BP, Lawson CL, Peisach E, Prlic A, Rose PW, Shao C, Yang H, Young J, Zardecki C (2013) Trendspotting in the protein data bank. FEBS Lett 587(8):1036–1045

23. Goodsell DS, Burley SK, Berman HM (2013) Revealing structural views of biology. Biopolymers 99(11):817–824

24. Wlodawer A, Miller M, Jaskolski M, Sathyanarayana BK, Baldwin E, Weber IT, Selk LM, Clawson L, Schneider J, Kent SB (1989) Conserved folding in retroviral proteases: crystal structure of a synthetic HIV-1 protease. Science 245(4918):616–621

25. Wensing AM, van Maarseveen NM, Nijhuis M (2010) Fifteen years of HIV protease inhibitors: raising the barrier to resistance. Antiviral Res 85(1):59–74

26. Cihlar T, Ray AS (2010) Nucleoside and nucleotide HIV reverse transcriptase inhibitors: 25 years after zidovudine. Antiviral Res 85(1):39–58

27. Yonath A (2005) Antibiotics targeting ribosomes: resistance, selectivity, synergism and cellular regulation. Annu Rev Biochem 74:649–679

28. Bulkley D, Innis CA, Blaha G, Steitz TA (2010) Revisiting the structures of several antibiotics bound to the bacterial ribosome. Proc Natl Acad Sci U S A 107(40):17158–17163

29. Rose PW, Bi C, Bluhm WF, Christie CH, Dimitropoulos D, Dutta S, Green RK, Goodsell DS, Prlic A, Quesada M, Quinn GB, Ramos AG, Westbrook JD, Young J, Zardecki C, Berman HM, Bourne PE (2013) The RCSB Protein Data Bank: new resources for research and education. Nucleic Acids Res 41(D1):D475–D482

30. Stierand K, Rarey M (2010) Drawing the PDB: protein – ligand complexes in two dimensions. Med Chem Lett 1:540–545

31. Knox C, Law V, Jewison T, Liu P, Ly S, Frolkis A, Pon A, Banco K, Mak C, Neveu V, Djoumbou Y, Eisner R, Guo AC, Wishart DS (2011) DrugBank 3.0: a comprehensive resource for 'omics' research on drugs. Nucleic Acids Res 39(Database issue):D1035–D1041

32. Young JY, Feng Z, Dimitropoulos D, Sala R, Westbrook J, Zhuravleva M, Shao C, Quesada M, Peisach E, Berman HM (2013) Chemical annotation of small and peptide-like molecules at the Protein Data Bank. Database 2013:bat079

33. Altschul SF, Gish W, Miller W, Myers EW, Lipman DJ (1990) Basic local alignment search tool. J Mol Biol 215:403–410

34. Hopper P, Harrison SC, Sauer RT (1984) Structure of tomato bushy stunt virus. V. Coat protein sequence determination and its structural implications. J Mol Biol 177(4):701–713

35. Schmeing TM, Voorhees RM, Kelley AC, Gao YG, Murphy FVT, Weir JR, Ramakrishnan V (2009) The crystal structure of the ribosome bound to EF-Tu and aminoacyl-tRNA. Science 326(5953):688–694

36. Voorhees RM, Weixlbaumer A, Loakes D, Kelley AC, Ramakrishnan V (2009) Insights into substrate stabilization from snapshots of the peptidyl transferase center of the intact 70S ribosome. Nat Struct Mol Biol 16(5):528–533

37. Gao YG, Selmer M, Dunham CM, Weixlbaumer A, Kelley AC, Ramakrishnan V (2009) The structure of the ribosome with elongation factor G trapped in the posttranslocational state. Science 326(5953):694–699

38. Nagar B, Hantschel O, Young MA, Scheffzek K, Veach D, Bornmann W, Clarkson B, Superti-Furga G, Kuriyan J (2003) Structural basis for the autoinhibition of c-Abl tyrosine kinase. Cell 112(6):859–871

Chapter 9
Small Molecule Crystal Structures in Drug Discovery

Colin Groom

Abstract The modern drug discovery scientist is bathed in structural information – structures of target proteins, candidate drug molecules and complexes of the two. Whilst the value of protein-ligand complexes is evident from many reports and reviews of the area, the use and impact of small molecule structures can, however, be more difficult to appreciate. Information about molecular conformations and interactions derived from these structures is such an integrated part of drug discovery that it is hard to tease out the particular value they bring. This chapter attempts to do just that.

9.1 Introduction

As long ago as 1929, just 16 years after the discovery of the technique of X-ray crystallography, the results of such experiments were being collected and shared [1]. Since then over 700,000 crystal structures of organic and metal-organic compounds have been published. This entire collection of structures in archived in the Cambridge Structural Database (CSD) [2].

The structures within the CSD represent a wide range of chemical functionality, including functional groups and substructures found in typical drug molecules. Moreover, the database contains the actual structures of many drug molecules [3].

9.2 Determination of Small Molecule Crystal Structures

Not only are the structures available for many drug molecules and molecules related to compounds of interest, but the determination of the structure of a small molecule is often facile. The primary hurdle to overcome remains the crystallisation of a molecule of interest. This must be available in a reasonable quantity and usually

C. Groom (✉)
Cambridge Crystallographic Data Centre, 12 Union Rd, Cambridge CB2 1EZ, UK
e-mail: groom@ccdc.cam.ac.uk

© Springer Science+Business Media Dordrecht 2015
G. Scapin et al. (eds.), *Multifaceted Roles of Crystallography in Modern Drug Discovery*, NATO Science for Peace and Security Series A: Chemistry and Biology, DOI 10.1007/978-94-017-9719-1_9

with a high level of purity in order to allow the generation of diffraction quality samples, typically a single crystal larger than 0.1 mm in each dimension. All drug discovery organisations, whether industrial or academic have access to small molecule crystallographic services, either internally or via collaboration.

The improvement in both experimental and computational powder diffraction techniques now also means that structures of drug-like molecules can often be solved without requiring single crystals [4]. This is particularly aided by use of geometrical information from existing structures [5].

9.3   Molecular Geometry and Conformation

The most common use of small molecule structures is in understanding molecular geometry and conformation. Since the 1970s bond lengths and valence angles for a huge range of chemical functionality have been available [6]. As the number of structures available has grown, it is now the norm to find not only this information for compounds containing functional groups of interest, but information regarding conformational preferences is now available for most drug substructures [7].

An understanding of conformational preferences is a prerequisite for compound design. Not only is it essential to understand which structural modifications to a lead compound may be appropriate, it is also key to understanding and achieving a high level of biological activity. As the energy penalty incurred by a molecule binding to its biological target in a suboptimal conformation will inevitably result in low affinity optimisation of conformation is a key element of drug design.

This has been exemplified in the design of inhibitors of the enzyme inosine monophosphate dehydrogenase [8, 9]. In order to bind to the target, inhibitors related to mycophenolic acid require a precise positioning of a carboxylic acid group with respect to a bicyclic group of the molecule. This can be achieved with an alkyl linker, with a central torsion angle of 110°. However, as seen in Fig. 9.1, the preferred torsion angle for this bond would be around 180°. Introducing an allylic bond into this system results in a preferred torsion of around 120°, closer to the desired value, and results in an increase in inhibition, as measured by IC_{50}, from 1.0 μM to 0.14 μM. Methylation of the carbon at one end of this bond results in a preferred torsion close to that of the required bioactive conformation, further increasing the potency to 0.02 μM. It is important to note that this improvement in binding is not primarily driven by improved interactions between inhibitors and the target protein, but by ensuring that the bound conformation is close to an energy minimum for the molecule.

The conformations of a specific molecule found in a small molecule crystal structure and that of the corresponding molecule bound to a protein may differ. However, only in very exceptional cases will these molecules not be in a low energy conformation. Therefore, analysis of the conformational preferences of molecules in small molecule structures is entirely relevant to protein-bound molecules. Indeed, where molecules are reported to adopt a protein–bound conformation that one would

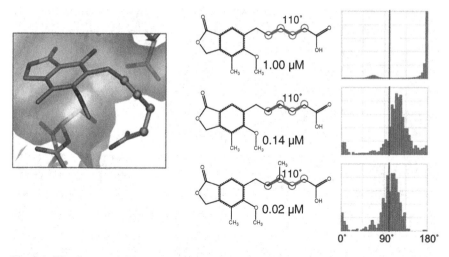

Fig. 9.1 The first panel illustrates the three dimensional conformation of mycophenolic acid bound to IMPDH. Associated with each compound structure is a histogram, describing the torsion angle distribution of molecules in the CSD with a corresponding structure. The torsion angle required for binding is highlighted by a *vertical line*. As an allyl bond is introduced, followed by methylations, this torsion angle distribution changes, until it closely matches to the desired value

not expect from an analysis of small molecule structures, it is usually due to an error in the interpretation of the electron density of the bound ligand [10]. It is worth noting, that information from small molecule structures is now a key part of the validation processes of the partners of the wwPDB involved in populating the Protein Data Bank. Analysis of small molecule structures is, without doubt, the most appropriate route to generate ligand restraint dictionaries for use by macromolecular crystallographers.

9.4 Molecular Interactions

The short range interactions between ligands and proteins are simply a subset of those that are observed in small molecule crystals – the collection of small molecule structures the community has generated incorporates all of the functionality seen in proteins. It is also probably reasonable to assume that the environment a molecule finds itself in inside a small molecule crystal is, in general terms, not unlike that inside a protein molecule. We can, therefore, use knowledge derived from small molecule structures to inform us as to how molecules may interact with a protein.

Software systems analogous to those described above for molecular geometry have also been developed for molecular interactions, based on small molecule crystal structures [11], protein ligand complexes [12] and both combined [13]. We

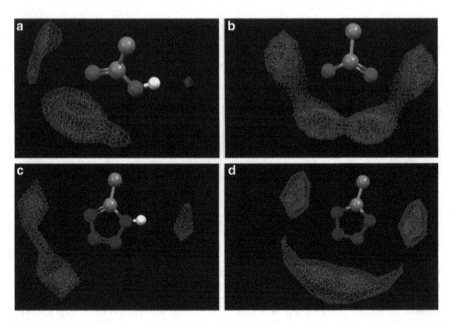

Fig. 9.2 Distributions of N-H and O-H donors around (**a**) –COOH and (**b**) –COO– groups, compared with distributions of N-H and O-H donors around (**c**) 1H-tetrazole and (**d**) tetrazolate moieties

can, therefore, optimise protein-ligand binding by optimising interactions to match the nature of those we observe in small molecule structures.

Of course, the flexibility of proteins can make this challenging, but nevertheless, small molecule structures are a perfect source of isosteric groups. This can be exemplified by acid functionality [14]. There are cases, for example, where a molecule may contain a carboxylic acid, and for various reasons, a replacement group is desired. A comparison of the interaction maps of carboxylic acids and other groups can be made. This reveals that a tetrazole moiety has a similar interaction map, suggesting it as a possible replacement, as seen in Fig. 9.2. This is a trivial example, known to every medicinal chemist, but the principle can be extended to groups with less obvious similarity.

Knowledge bases of molecular interactions also underpin pharmacophore methods [15] and protein-ligand interaction scoring functions, which are fundamental to molecular docking approaches [16].

9.5 Molecular Properties

We've seen that small molecule crystal structures can inform us of the conformational and interaction preferences of a molecule. They can, however, also teach us about fundamental materials properties, for example solubility. This is nicely

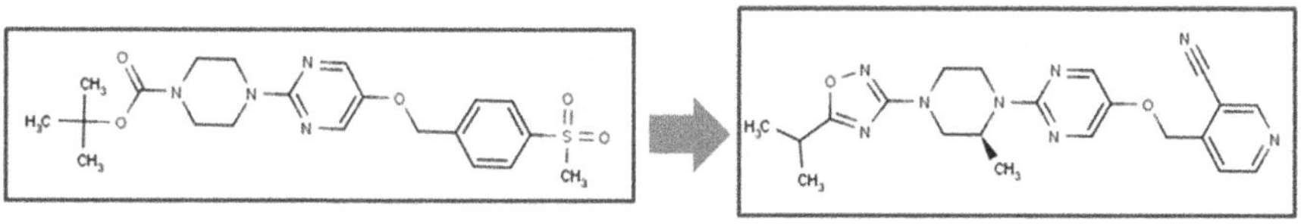

GPR119 EC$_{50}$	65 nM	GPR119 EC$_{50}$	6 nM
Solubility	0.03 µM	Solubility	6 µM
LogD$_{7.4}$	3.2	LogD$_{7.4}$	3.3

Fig. 9.3 Improvements to GPR119 agonists, following a strategy of introducing modifications that reduced the 'self-complementarity' of the molecule

exemplified by work reported by AstraZeneca [17]. Faced with an agonist of the receptor GPR119, with attractive activity but poor solubility, small molecule crystallography was used to understand the key interactions in a crystal of the molecule. Making modifications that omitted these interactions delivered a molecule that retained potency but was also 200 fold more soluble, as shown in Fig. 9.3.

9.6 Solid Form Properties

The reach of small molecule crystallography extends beyond the realm of work described as drug discovery and into that of drug development. Here, the molecular structure of a molecule is fixed and the challenge is to develop a form of the active pharmaceutical ingredient suitable for manufacture, clinical trials and, ultimately, sale.

9.7 Polymorphism

It is vital that the production of a drug form is a repeatable, well-understood process. One area of particular focus is ensuring that a single polymorph of a compound is produced. Many experimental protocols exist to establish relative polymorph stability, however, supplementing these using structural informatics, illustrated in Fig. 9.4, can establish a wider perspective, e.g. whether sufficient experimental screening has been carried out. One particularly powerful technique is to assess the hydrogen bonding patterns in a lattice. Poor satisfaction of hydrogen bonding potential, and the presence of unusual hydrogen bonds may be indicative of a metastable lattice [18]. Where hydrogen bonding is not the dominant feature of molecular association, a critical analysis of all interactions might indicate the likely stability of a lattice [19].

a

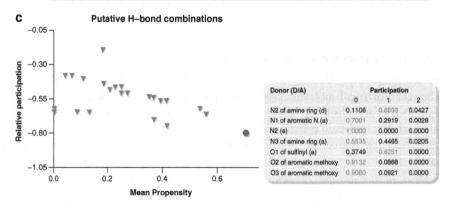

b

Donor (D/A)	Acceptor	Propensity	Intermolecular interaction observed	Bounds +/-
N2 of amine ring	O1 Sulfinyl	0.74	Yes	(0.66, 0.81)
N2 of amine ring	N3 of amine ring	0.39		(0.26, 0.55)
N2 of amine ring	N1 of aromatic N	0.36		(0.23, 0.51)
N2 of amine ring	O3 aromatic methoxy	0.15		(0.09, 0.26)
N2 of amine ring	O2 Aromatic methoxy	0.11		(0.06, 0.19)

c

Putative H–bond combinations

Donor (D/A)	Participation		
	0	1	2
N2 of amine ring (d)	0.1106	0.8893	0.0427
N1 of aromatic N (a)	0.7081	0.2919	0.0028
N2 (a)	1.0000	0.0000	0.0000
N3 of amine ring (a)	0.5535	0.4465	0.0205
O1 of sulfinyl (a)	0.3749	0.6251	0.0000
O2 of aromatic methoxy	0.9132	0.0868	0.0000
O3 of aromatic methoxy	0.9080	0.0921	0.0000

Fig. 9.4 (**a**) The 2D structure of the drug omeprazole. (**b**) A calculation of the expectation of seeing particular hydrogen bonds in the crystal structure of this molecule (the hydrogen bond propensity). (**c**) A plot of the hydrogen bond propensity against the expected number of hydrogen bonds (the participation). The *triangles* represent all possible hydrogen bonding arrangements in the crystal lattice. The *circle* represents the observed crystal structure of omeprazole

9.8 Co-crystals

Drug products are not always created from crystals of the active pharmaceutical ingredient alone. Co-crystals – crystalline forms composed from two or more components with a particular stoichiometry – are often used. Within this broad defi-nition, salts (where some components are charged), hydrates (where one component is water) and solvates (where one component is a solvent) are also frequently used.

Co-crystallisation has shown promise in the tuning of a range of physical properties including dissolution rate [20], compressibility [21] and physical stability [22].

9.9   Crystal Structure Prediction

Of course being able to predict the crystal structure of a molecule would allow one to attempt to predict the lattice properties prior to a molecule even being synthesised. This is non-trivial as it involves understanding the fine balance between competing molecular interactions and geometrical optimisation. Often, many plausible lattices can be generated, each separated by a small energetic difference. Despite these difficulties, there has been substantial progress. Whilst being neither routine, nor giving highly reliable results, the structures of many typical small molecule therapeutics can be predicted [23].

9.10   Crystal Morphology

The morphology (habit, or shape) of crystalline particles can also have a large impact on the production of a drug substance. Again, attempts have to been to predict the size and shape of crystals [24].

9.11   Discussion

Small molecule crystal structures are of tremendous value in drug discovery. An understanding of molecular conformation and interactions is fundamental to the optimisation of protein-ligand interactions. Even when an active molecule with appropriate properties is identified, much further work remains in order to turn such a molecule into a drug. Again, small molecule crystals structures are of unparalleled value in this process.

References

1. Ewald PP, Hermann C (1929) Strukturbericht 1913–1928. Akademische Verlagsgesellschaft, Leipzig
2. Allen FH (2002) The Cambridge structural database: a quarter of a million crystal structures and rising. Acta Crystallogr B 58:380–388
3. McGrath NA, Brichacek M, Njardarson JT (2010) A graphical journey of innovative organic architectures that have improved our lives. J Chem Educ 87:1348–1349

4. David WIF, Shankland K, Shankland N (1998) Routine determination of molecular crystal structures from powder diffraction data. Chem Commun 8:931–932
5. David WIF, Shankland K, van de Streek J et al (2006) DASH: a program for crystal structure determination from powder diffraction data. J Appl Cryst 39:910–915
6. Kennard O, Watson DG et al (1970–1986) Molecular structures and dimensions, bibliographic volumes 1–16 and volume A1 D. Reidel Publishing Company, Dordrecht
7. Bruno IJ, Cole JC, Kessler M et al (2004) Retrieval of crystallographically-derived molecular geometry information. J Chem Inf Comput Sci 44:2133–2144
8. Brameld KA, Kuhn B, Reuter DC et al (2008) Small molecule conformational preferences derived from crystal structure data. A medicinal chemistry focused analysis. J Chem Inf Model 48:1–24
9. Bissantz C, Kuhn B, Stahl M (2010) A medicinal chemist's guide to molecular interactions. J Med Chem 53:5061–5084
10. Liebeschuetz J, Hennemann J, Olsson T et al (2012) The good, the bad and the twisted: a survey of ligand geometry in protein crystal structures. J Comput Aided Mol Des 26:169–183
11. Bruno IJ, Cole JC, Lommerse JPM et al (1997) Isostar: a library of information about nonbonded interactions. J Comput Aided Mol Des 11:525–537
12. Verdonk ML, Cole JC, Taylor R (1999) SuperStar: a knowledge-based approach for identifying interaction sites in proteins. J Mol Biol 289:1093–1108
13. Nissink JWM, Taylor R (2004) Combined use of physiochemical data and small-molecule crystallographic contact propensities to predict interactions in protein binding sites. Org Biomol Chem 2:3238–3249
14. Allen FH, Groom CR, Liebeschuetz JW et al (2012) The hydrogen bond environments of 1 H-tetrazole and tetrazolate rings: the structural basis for tetrazole–carboxylic acid bioisosterism. J Chem Inf Model 52:857–866
15. Leach AR, Gillet VJ, Lewis RA et al (2010) Three-dimensional pharmacophore methods in drug discovery. J Med Chem 53:539–558
16. Velec HFG, Gohlke H, Klebe G (2005) DrugScore(CSD) knowledge-based scoring function derived from small molecule crystal data with superior recognition rate of near-native ligand poses and better affinity prediction. J Med Chem 48:6296–6303
17. Scott JS, Birch AM, Brocklehurst KJ et al (2012) Use of small-molecule crystal structures to address solubility in a novel series of G protein coupled receptor 119 agonists: optimization of a lead and in vivo evaluation. J Med Chem 55:5361–5379
18. Galek PT, Allen FH, Fábián L et al (2009) Knowledge-based H-bond prediction to aid experimental polymorph screening. CrystEngComm 11:2634–2639
19. Wood PA, TSG Olsson TSG, Cole JC (2012) Evaluation of molecular crystal structures using Full Interaction Maps. CrystEngComm 15:65–72
20. Žegarac M, Lekšić E, Šket P et al (2014) A sildenafil cocrystal based on acetylsalicylic acid exhibits an enhanced intrinsic dissolution rate. CrystEngComm 16:32–35
21. Karki S, Friščić T, Fábián L et al (2009) Improving mechanical properties of crystalline solids by cocrystal formation: new compressible forms of paracetamol. Adv Mater 21:3905–3909
22. Trask AV, Motherwell WDS, Jones W (2006) Physical stability enhancement of theophylline via cocrystallization. Int J Pharm 320:114–123
23. Bardwell DA, Adjiman CS, Arnautova YA et al (2011) Towards crystal structure prediction of complex organic compounds-a report on the fifth blind test. Acta Crystallogr B 67(6):535–551
24. Zhang Y, Doherty MF (2004) Simultaneous prediction of crystal shape and size for solution crystallization. AICHE J 50(9):2101–2112

Chapter 10
Protein Aggregation and Its Prediction

Ricardo Graña-Montes and Salvador Ventura

Abstract The presence of protein aggregates in tissues is the hallmark of more than 40 different human disorders, from neurodegenerative diseases to systemic and localized amyloidosis. On the other hand recombinant protein production is an essential tool for the biotechnology industry and supports expanding areas of basic and biomedical research, including structural genomics and proteomics. However, many recombinant polypeptides, specially those of human origin, undergo irregular or incomplete folding processes, when produced in heterologous hosts, that usually result in their accumulation as insoluble aggregates which are known as inclusion bodies, and resemble those formed in conformational disorders. Consequently, many biologically relevant protein-based drugs are excluded from the market simply because they cannot be harvested in their native form at economically convenient yields. As discussed herein, the biomedical and biotechnological relevance of these two different, but mechanistically related, problems has pushed the study of protein aggregation to evolve from a barely neglected area of protein chemistry to a highly dynamic research field nowadays.

10.1 Introduction

The folding of a polypeptide chain into its native structure is a complex process; thus, errors along folding may occur and drive proteins to incorrectly folded or misfolded states, which may still possess certain stability in the physiological environment and, therefore, start to accumulate. The inability of a protein to achieve or maintain its native conformation, resulting in the formation of different types of aggregates, is associated to an increasing number of human pathologies ranging from neurodegenerative disorders, such as Alzheimer's and Parkinson's diseases, to different amyloidoses, characterized by the formation of large proteinaceous deposits, or even diabetes mellitus type II and certain types of cancer [1–3].

R. Graña-Montes • S. Ventura (✉)
Institut de Biotecnologia i Biomedicina, Universitat Autònoma de Barcelona, 08193 Bellaterra, Barcelona, Spain
e-mail: Salvador.Ventura@uab.es

© Springer Science+Business Media Dordrecht 2015   115
G. Scapin et al. (eds.), *Multifaceted Roles of Crystallography in Modern Drug Discovery*, NATO Science for Peace and Security Series A: Chemistry and Biology, DOI 10.1007/978-94-017-9719-1_10

The first references we have regarding the presence of anomalous deposits in human tissues that might be associated to amyloid-like protein aggregates come from seventeenth century autopsy reports [4]. Later nineteenth century medical reports tell us about unusual inclusions in different organs, mainly in the liver and the spleen, of pale-colored and dense substances, which possibly also corresponded to accumulations of amyloid proteins [5]. In 1854, Rudolph Virchow was the first who employed the term "amyloid" to name cerebral structures with abnormal appearance which were stained with iodine—thus he considered them related to starch—, and were similar to those previously mentioned inclusions. Although we nowadays know Virchow was actually observing *corpora amylacea*, which are mostly composed of glycosaminoglycans [6, 7], the chemical analysis of "amyloid" materials made first by George Budd and later by Carl Friedreich and August Kekulé lead them to conclude these substances were mostly "albuminous", meaning that they posses a proteinaceous nature [7].

Although the polypeptidic nature of amyloid substances had already been established in the nineteenth century, and despite of the progress made during the twentieth century towards the classification of the different diagnosed amyloidoses, it was long believed that amyloid was rather a concrete substance of possible unspecific degenerative origin [8]. It was not until the 1970s that the biochemical characterization of different types of amyloids allowed to establish that each of them was primarily composed of a specific kind of protein or variants of the same protein. Nowadays, the term "amyloid" is employed to denominate a type of essentially proteinaceous, extracellular aggregates, which are associated to more than 40 human pathologies; although amyloids with a specific biological function have been described in several species, including humans [9].

Amyloid aggregates are characterized by a particular fibrilar morphology under transmission electron microscopy (TEM), which is highly ordered, compact, stable and unbranched. This type of structures are denominated amyloid fibrils, their diameters range within tens of nanometers and their longitude can reach several micrometers; mature fibrils can further associate laterally to form fibers. Amyloid fibrils have been shown to share a common molecular architecture composed of the cross-ß supersecondary structure, where parallel β-sheets extend with their strands facing to each other and perpendicular to the fibril axis; such a conformation possesses a characteristic X-ray diffraction pattern [10]. This particular structure provides amyloid fibrils with the capability of binding certain chemical compounds like Thioflavin-T (Th-T) and Congo Red (CR), whose spectral properties change upon binding to amyloid fibrils, therefore serving as probes to test the amyloid-like nature of protein aggregates [11]. Aside from TEM, fibril morphology can also be detected employing atomic force microscopy (AFM), and the conformational conversion from the native conformation may be followed with Fourier-transform infrared spectroscopy (FTIR) by monitoring the appearance of the characteristic intermolecular β-sheet band at $\approx 1,625$ cm^{-1}, or with circular dichroism (CD) by following changes in the typical negative β-sheet signal at ≈ 217 nm [12, 13].

The polypeptides involved in the formation of abnormal amyloid-like protein deposits are neither related in terms of sequence, nor in native conformation—some of them being predominantly unstructured (i.e. the amyloid β peptide and α-synuclein) while others are compact globular proteins in their native state (i.e. transthyretin, superoxide dismutase 1 and β2-microglobulin) [2]. On the other hand, aggregation into amyloid-like fibrils similar to those found in protein deposits is not restricted to a discrete number of polypeptides related to certain human pathologies, but has been shown or induced for a large number of proteins, from different organisms, many of them lacking any known association to disease [14–17]. Moreover, cross-β structure has also been reported in macroscopically non-fibrilar, apparently amorphous, aggregates [18]. These findings have led to the consideration that the ability to adopt the cross-β supersecondary conformation, would constitute an intrinsic property of virtually any polypeptide [19] since backbone-mediated interactions are the strongest contributors towards the acquisition of such conformations [20].

10.2   Results and Discussion

The experimental analysis of amyloidogenic proteins and peptides has revealed that changes on its amino acid sequence lead to dramatic changes on its tendency to aggregate [21–23]. These results show how the primary structure of proteins plays an extremely relevant role in determining the tendency of polypeptides to form insoluble deposits; this would arise from the impact over such propensity of inherent physical-chemical properties of amino acids such as hydrophobicity, the structural suitability to adopt β-conformation or its mean charge [23]. Specifically, the study of polypeptides able to form amyloid-like fibrils but lacking any defined three-dimensional structure on its physiological context, such as the intensively explored Aβ peptides related to the Alzheimer's disease, has allowed to decouple the determinants of protein aggregation from those of the folding of globular proteins, whose driving forces substantially overlap [24].

10.2.1   Intrinsic Determinants of Protein Aggregation

Hydrophobicity appears to be one of the major determinants promoting aggregation. It has been shown for many proteins or peptides, that mutations substituting polar by non polar residues tend to increase their aggregation propensity and or deposition rate [24]. However, it has been shown that hydrophobicity does not suffice by itself to rationalize the outcome of amino acid replacements on the aggregation potential. Accordingly, attempts to predict protein aggregation propensity solely on the basis of side-chain hydrophobicity have failed [21].

Another relevant intrinsic property of polypeptides is its propensity to adopt specific secondary structure motifs. Consistent with the observation that most protein aggregates share the cross-β supersecondary structure, it has been observed that aggregation is favoured by aminoacids with a higher propensity to adopt β-sheet conformation [25, 26]. Moreover, pre-existing β-strands in protein structures also contribute to the tendency of polypeptides to aggregate [17]. Consequently, aminoacids disfavouring β-conformation, such as Pro and Gly, have been reported to largely disrupt the aggregation propensity of sequence streches [27].

As a physical-chemical property that can be globally regarded as opposed to hydrophobicity, amino acid charge is known to prevent or disrupt protein deposition [21]. As well, the net charge of the whole polypeptide also influences aggregation [28]; the higher a protein net charge, the greater the repulsion between individual protein molecules and the lower their chances to establish intermolecular contacts.

The aforementioned factors may be regarded as intrinsic determinants arising from individual properties of amino acids. Nonetheless, the linear combination of their properties along the primary structure has a cooperative impact over the aggregation propensity. It has been observed that the consecutive occurrence of three or more hydrophobic residues is clearly disfavored in nature [29]. In a similar way, the combinatorial design of amyloidogenic proteins has shown how polypeptidic patterns alternating apolar and polar aminoacids favor amyloid formation [30]. Remarkably, it has been shown that these patterns are less frequent in natural proteins than it would be expected by random chance [31].

The failure of protein aggregates to adopt regular macroscopic assemblies hampered, for many years, the possibility of obtaining a detailed description of the structure of amyloids at an atomic level. Fortunately, the development of new techniques, such as solid state NMR (ss NMR) [32] or microcrystallization of amyloidogenic peptides [33] has allowed to unveil the molecular detail of amyloid formation for certain proteins and short peptides. The solved structures provide an outstanding framework to rationalize the intrinsic determinants of protein aggregation. Many of the solved structures correspond to an extended β-sheet whose β-strands run perpendicular to the axis of the fibril. In these β-sheets, hydrophobic residues are protected from the solvent by establishing interactions with other apolar residues of β-strands in the opposite β-sheet, while polar residues are exposed to the solvent The geometry of the β-conformation allows the side chains of contiguous residues to point in opposite senses, so this explains how alternation of non polar and polar residues in the primary structure facilitates amyloid formation.

Finally, the sequence of a polypeptide determines its folding to a defined three-dimensional structure, which, under physiological conditions, would correspond to its native structure. Therefore, mutations in the primary structure may affect both the structure and the stability of a protein. These variables affect the propensity and the rate to which a polypeptide may aggregate; therefore native conformation and its stability may be considered intrinsic determinants of protein aggregation. Furthermore, these parameters are also intimately related to the mechanisms by which polypeptides aggregate.

10.2.2 Environmental Determinants of Protein Aggregation

The extrinsic determinants of protein aggregation refer to a set of variables defining the environment of the polypeptide chain, which can affect the tendency of a protein to form deposits since they are able to modulate the intrinsic factors governing aggregation. The most relevant extrinsic determinants are the pH and the ionic strength of the solution, together with the temperature of the system [34]. These variables may affect both the kinetics and thermodynamics of aggregation into amyloid-like structures, and can, subsequently, influence the assembly and the macroscopic structure of the aggregated species, thus being important determinants of the polymorphism the aggregates of a given protein sequence may present.

The pH directly influences the protonation state of amino acid sidechains and, accordingly, modulates physical-chemical properties such as its polarity and net charge, which, as described above, posses a strong relevance in the aggregation propensity of polypeptides. In the same way, pH also influences the net charge of the protein, thus modulating electrostatic repulsion between individual molecules and, subsequently, the probability of establishment of the intermolecular interactions required for the formation of amyloid-like structures. Ionic strength has a similar role in modulating aggregation since its increase allows to shield amino acid sidechain charges and, therefore, to decrease the repulsion between polypeptide molecules [35].

As an intrinsic determinant, the conformational stability of a polypeptide may also be altered by external factors. For those proteins which adopt a defined three-dimensional structure in its physiological context, variables such as temperature or pH my alter the network of interactions stabilizing their native conformations, thus allowing polypeptides to populate partially or globally unfolded states from where aggregation might take place more easily (Fig. 10.1).

10.2.3 Specific Regions Determining Protein Aggregation

The intrinsic determinants described before inform us about specific properties of the amino acid side chain either favoring or disfavoring protein aggregation. The linear combination of such properties within the primary structure plays a major role in protein aggregation. However, it has been observed that not all the polypeptide sequence has the same importance in defining its propensity to aggregate. There exist small amino acid stretches within protein sequences, which promote and guide protein aggregation into amyloid-like structures [36]. These short fragments, generally referred to as aggregation-prone regions (APRs) or "hot-spots", are characterized by an enrichment in hydrophobic, both aliphatic (Val, Leu, Ile) and aromatic (Phe, Trp, Tyr), residues [37]. The analysis of the structural models

Monomeric globular proteins   Natively disordered proteins

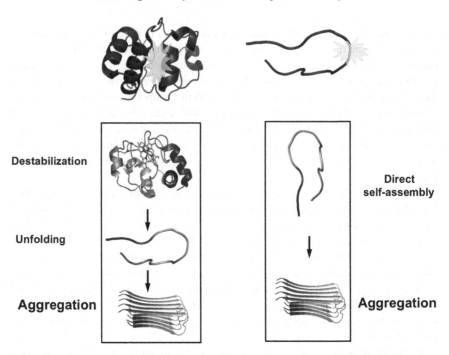

Fig. 10.1 Aggregation of globular and intrinsically disordered proteins. In globular proteins aggregation-prone regions (APRs) are usually protected from the solvent. Intrinsic or extrinsic factors destabilizing thermodynamically or kinetically the protein might promote local or global unfolding resulting in the transient exposure of APRs, which might lead to aggregation. In intrinsically disorder proteins (IDPs) APRs would be readily exposed to solvent promoting fast self-assembly and aggregation. Therefore, IDPs tend to be depleted in APRs in their sequences

for some amyloids [38] also allow to rationalize the reason why APRs direct the formation of amyloid-like structures, since the cross-β arrangement in the core of amyloid fibrils does only strictly require the minimum participation of a single β-strand per molecule, and the rest of the polypeptide may remain exposed to the solvent even when "attached" along the fibril. APRs are usually located within or substantially take part in the hydrophobic cores of the native state of proteins [39] and also frequently map to protein-protein interaction surfaces of protein adopting a stable quaternary structure [40, 41], which prevents APRs from establishing aberrant intermolecular contacts.

The examination of APRs in the context of entire proteins has also revealed how these stretches are usually flanked by charged residues (Asp, Glu, Lys, Arg), whose function would be to hamper intermolecular interactions between APRs in the event they become exposed by providing repulsive charge, or by residues acting as β-sheet breakers, like Pro [37].

10.2.4 Mechanisms of Protein Aggregation

The strikingly divergent tertiary and quaternary structures adopted by those proteins able to form amyloids indicates that no general model would allow to account for the specific mechanism of conversion. Several models have been proposed to explain the conformational conversion required for amyloid-like aggregation based on both the structure and the conformational stability of polypeptides [33].

A first model relies on the observation that many amyloids are composed of proteins or fragments of proteins that lack a defined three-dimensional structure in its physiological environment, like amyloid β or α-synuclein. According to this "Intrinsically Unstructured" model, aggregation-prone polypeptides without a defined three-dimensional conformation can establish intermolecular interactions in their native (unstructured) conformation (Fig. 10.1). However, such consideration does not imply this class of proteins can readily form amyloid-like structures, they might also require a transition to a conformation compatible with amyloid structure [42], as this is supported by the fact a lag phase is also observed in the aggregation kinetics of this kind of proteins.

The establishment of intermolecular interactions between APRs within intrinsically unstructured conformations can be rationalized in a simple manner, but the way APRs protected by the native state of globular proteins get to interact to form amyloid-like structures is not straightforward. A different mechanism to explain the conversion of globular proteins into amyloid-like structures, the "Refolding" model, proposes substantial unfolding of the native conformation is required for APRs to become exposed and being able to establish effective interactions leading to conformational conversion into the cross-β fold. The substantial perturbation of the native state required to populate such largely unfolded conformations, as those observed at the starting point of the aggregation into amyloid-like fibrils induced for proteins like the SH3 domain or myoglobin [14, 43], would arise from the impact in the global stability of the protein of mutations or changes in the environmental conditions (Fig. 10.1).

However, overcoming the usually large unfolding barrier, as required by the "Refolding" model, does not seem feasible for a vast majority of aggregation-prone globular proteins simply by means of single substitutions or normal changes in its physiological environment. Indeed, aggregation into amyloid-like fibrils has been reported for proteins departing from native-like conformations, like acylphosphatase and lysozyme [44], which in some cases are even able to retain a certain functional activity [45, 46]. Another mechanism of conformational conversion, the "Gain-of-Interaction" model, has been proposed to rationalize aggregation into amyloid-like structures without requiring crossing of the unfolding barrier. According to this model, local conformational perturbations of the native state would be sufficient to expose previously protected APRs in a way they can now establish non-functional intermolecular contacts with other polypeptide units displaying similarly exposed stretches. These locally disordered states can be reached directly through fluctuations of the native conformation, while the remainder of the native structure may well be left unaltered and the protein molecule may even retain its physiological activity to a certain extent. The precise mechanism through which this gain-of-interaction is achieved depends on several factors such as the fold and size of the protein, the location of APRs within the native conformation and the quaternary structure of the polypeptide. Three different subtypes among the "Gain-of-Interaction" model have been proposed to account for such variables.

The simplest mode of gain-of-interaction implies exposed APRs directly interact with each other to form the cross-β spine of the fibril, while the remainder of the polypeptide is left "hanging" attached to this fibril spine. This model requires a certain self-complementarity between the sidechains of the polypeptide APRs which will constitute the future strands of the extended β-sheet that defines the spine of the fibril. Extension of the fibril results from the successive stacking of APRs to form the β-sheet through the establishment of backbone hydrogen bonds. The analysis of the X-ray structures derived from microcrystals of fibril-forming peptides has allowed to define the geometrical requirements for APRs to build cross-β spines following a complementarity scheme that has been termed as steric zipper [47].

Another mode through which polypeptides may undergo gain-of-interaction consists in the rearrangement of APRs to the surface of the protein, so a novel homo-oligomerization surface is generated, therefore allowing direct stacking of protein monomers as a mechanism of fibril extension [48]. This model of conformational conversion has been proposed for the aggregation of all-β proteins with a native oligomeric quaternary structure, such as superoxide dismutase 1 (SOD1) or transthyretin (TTR) [48, 49]. These proteins posses a β-sandwich fold and display protein-protein interaction surfaces that allow them to form native, soluble oligomeric structures. In the conversion into amyloid-like structures of both proteins, either partial or global preservation of native dimerization surfaces has been proposed, but the emergence of non-functional interaction surfaces in the opposite side of the molecule, able to establish contacts with identical regions of neighbouring polypeptides, allows protein oligomerization to extend linearly into amyloid-like fibrilar structures where the strands of the mostly native β-sandwich are disposed perpendicular to the fibril axis, thus mimicking the cross-β assembly.

Finally, a third alternative mode for proteins to experience gain-of-function is 3D domain swapping where at least two identical protein entities exchange part of their structural elements to yield stable oligomeric species, as it has been show for cystatin C [50]. Extension of the oligomer size may be compatible with growth into an amyloid-like fibril when the swapping mechanism results in the formation of β-conformation able to dock in a cross-β spine fashion, or when it involves the successive structural exchange between adjacent polypeptide units [51].

10.2.5   Prediction of Protein Aggregation-Prone Regions and Protein Aggregation Propensities from the Primary Sequence

As introduced previously, the primary structure of a polypeptide strongly influences its aggregation propensity and point mutations may have a huge impact on protein solubility [23]. Known intrinsic properties of amino acids and polypeptides that can affect aggregation include charge, hydrophobicity. Furthermore, APRs typically consist of 5–15 amino acids segments which can nucleate the aggregation of the entire protein. The knowledge accumulated in the past 10 years on protein deposition processes has facilitated the flourishing of algorithms able to predict and characterize the aggregation propensity of proteins starting from its primary sequence. To develop these approaches, researchers have employed a high diversity of sources and premises coming from in vitro or in vivo experimental data, structural parameters or biophysical properties of polypeptides. These computational approaches have proved to be remarkably helpful in the design of strategies to control protein deposition events. The easy access to these bioinformatic tools and their overall accuracy has resulted in a significant number of published works coming from different research areas, that exploit these predictive tools to gain insights on the self-assembly properties of structurally and sequentially unrelated proteins or protein sets [52, 53]. These algorithms include: AGGRESCAN (http://bioinf.uab.es/aggrescan/) [54] based on an experimental scale of aggregation propensity determined in vivo, FoldAmyloid (http://antares.protres.ru/fold-amyloid/) [55], which exploits the higher burial and hydrogen bonding capability observed in amyloid stretches for prediction, PASTA (http://biocomp.bio.unipd.it/pasta/) [56], which uses a non-redundant set of known globular structures to statistically derive pairing energies for the amino acids that form contacts between adjacent β-strands in the β-sheets of the amyloid-like spine, Zyggregator (http://www-vendruscolo.ch.cam.ac.uk/zyggregator.php) [57] that uses a set of physico-chemical properties of amino acid residues such as hydrophobicity, charge, and the propensity to adopt α-helical or β-sheet conformations, Waltz (http://waltz.switchlab.org/) [58], which combines experimental data on the amyloid character of an hexapeptide database with physicochemical and structure-based parameters and ZipperDB (http://services.mbi.ucla.edu/zipperdb/) [59], which models sequences on top of the crystal structure of "cross-β spine". All these methods have been shown to provide accurate predictions and, in many cases, the putative protein aggregation-regions suggested by these conceptually different algorithms overlap significantly.

10.2.6 Prediction of Protein Aggregation-Prone Patches in 3D Structures

The characterization and control of aggregation has become a central concern in the field of protein therapeutics and biological formulation. Traditionally, this is addressed via wet experiments carried physically in the laboratory using trial and error approaches that are expensive, difficult to perform, and time consuming.

In this context, it is predicted that computational tools would result in millions of euros in savings downstream, thanks to the gain in efficiency attained when working with pre-selected, well-characterized proteins, especially at the early development and formulation stages. This emphasizes the need for new in silico tools than can be used to rank protein-based drugs, and help in the selection of the proper molecules for product development. Despite the above-described algorithms perform fairly well in detecting aggregation-nucleating sequences in proteins primary sequences, it is also true that many of these sequences are actually hidden whithin the inner core of the protein in its folded state, and would not contribute to aggregation unless the protein unfolds extensively. Such unfolding is not expected once the protein has been produced and purified in the native state. Therefore, the challenge is to predict the aggregation propensities of proteins in the context of their 3D structures.

A first clue to develop this kind of algorithms comes from the recent observation that certain amyloidogenic proteins involved in disease posses a globular quaternary structure or interact obligatorily with other proteins, and that the interfaces sustaining these oligomeric structures are aggregation-prone [41]. The spatial coincidence of interaction sites and aggregating regions suggests that the formation of functional complexes and the aggregation of their individual subunits might compete in the cell. Accordingly, single mutations affecting the complex interface structure or stability usually result in the formation of toxic aggregates. In fact, the overlap between productive and anomalous interaction sites is, again, not restricted to disease related proteins but rather a generic property of protein-protein interactions and, actually, aggregation propensity can be used to distinguish protein interfaces from their surfaces [40]. Therefore, it is inferred that algorithms able to detect aggregation-prone patches close to protein surfaces would allow identifying the regions through which protein therapeutics could interact abrerrantly during downstream processing or storage.

One algorithm suitable for this type of predictions is SAP (for Spatial Aggregation Propensity), developed and patented at the Massachusetts Institute of Technology and recently licensed by Accelrys [60]. SAP Identifies hot-spots for aggregation based on the dynamic exposure of spatially-adjacent hydrophobic amino acids and has been essentially applied to the redesign of therapeutic antibodies. It is likely that as, it happened before with prediction from protein sequences, in the next few years we foresee the advent of a myriad of algorithms to predict protein aggregation within 3D structures, thus opening new ways to tackle this recurrent issue in protein-based drug design projects, as well as to find therapeutic avenues for the remediation of the devastating diseases caused by protein aggregation.

References

1. Selkoe DJ (2003) Folding proteins in fatal ways. Nature 426(6968):900–904
2. Chiti F, Dobson CM (2006) Protein misfolding, functional amyloid, and human disease. Annu Rev Biochem 75:333–366
3. Invernizzi G, Papaleo E, Sabate R, Ventura S (2012) Protein aggregation: mechanisms and functional consequences. Int J Biochem Cell Biol 44(9):1541–1554
4. Kyle RA (2001) Amyloidosis: a convoluted story. Br J Haematol 114(3):529–538

5. Doyle L (1988) Lardaceous disease: some early reports by British authors (1722–1879). J R Soc Med 81(12):729–731
6. Sakai M, Austin J, Witmer F, Trueb L (1969) Studies of corpora amylacea. I. Isolation and preliminary characterization by chemical and histochemical techniques. Arch Neurol 21(5):526–544
7. Sipe JD, Cohen AS (2000) Review: history of the amyloid fibril. J Struct Biol 130(2–3):88–98
8. Westermark P (2005) Aspects on human amyloid forms and their fibril polypeptides. FEBS J 272(23):5942–5949
9. Fernandez-Busquets X, de Groot NS, Fernandez D, Ventura S (2008) Recent structural and computational insights into conformational diseases. Curr Med Chem 15(13):1336–1349
10. Sunde M, Blake CC (1998) From the globular to the fibrous state: protein structure and structural conversion in amyloid formation. Q Rev Biophys 31(1):1–39
11. Nilsson MR (2004) Techniques to study amyloid fibril formation in vitro. Methods 34(1):151-160
12. Pallares I, Vendrell J, Aviles FX, Ventura S (2004) Amyloid fibril formation by a partially structured intermediate state of alpha-chymotrypsin. J Mol Biol 342:321–331
13. Sabate R, Ventura S (2013) Cross-beta-sheet supersecondary structure in amyloid folds: techniques for detection and characterization. Methods Mol Biol 932:237–257
14. Guijarro JI, Sunde M, Jones JA, Campbell ID, Dobson CM (1998) Amyloid fibril formation by an SH3 domain. Proc Natl Acad Sci U S A 95(8):4224–4228
15. Ventura S, Lacroix E, Serrano L (2002) Insights into the origin of the tendency of the PI3-SH3 domain to form amyloid fibrils. J Mol Biol 322:1147–1158
16. Sabate R, Espargaro A, Grana-Montes R, Reverter D, Ventura S (2012) Native structure protects SUMO proteins from aggregation into amyloid fibrils. Biomacromolecules 13(6):1916-1926
17. Castillo V, Chiti F, Ventura S (2013) The N-terminal helix controls the transition between the soluble and amyloid states of an FF domain. PLoS One 8(3):e58297
18. Wasmer C, Benkemoun L, Sabate R, Steinmetz MO, Coulary-Salin B, Wang L, Riek R, Saupe SJ, Meier BH (2009) Solid-state NMR spectroscopy reveals that E. coli inclusion bodies of HET-s(218–289) are amyloids. Angew Chem Int Ed Engl 48(26):4858–4860
19. Fandrich M, Fletcher MA, Dobson CM (2001) Amyloid fibrils from muscle myoglobin. Nature 410:165–166
20. Knowles TP, Vendruscolo M, Dobson CM (2014) The amyloid state and its association with protein misfolding diseases. Nat Rev Mol Cell Biol 15(6):384–396
21. Wurth C, Guimard NK, Hecht MH (2002) Mutations that reduce aggregation of the Alzheimer's Abeta42 peptide: an unbiased search for the sequence determinants of Abeta amyloidogenesis. J Mol Biol 319:1279–1290
22. de Groot NS, Aviles FX, Vendrell J, Ventura S (2006) Mutagenesis of the central hydrophobic cluster in Abeta42 Alzheimer's peptide. Side-chain properties correlate with aggregation propensities. FEBS J 273(3):658–668
23. Chiti F, Stefani M, Taddei N, Ramponi G, Dobson CM (2003) Rationalization of the effects of mutations on peptide and protein aggregation rates. Nature 424(6950):805–808
24. Jahn TR, Radford SE (2008) Folding versus aggregation: polypeptide conformations on competing pathways. Arch Biochem Biophys 469(1):100–117
25. Chiti F, Taddei N, Baroni F, Capanni C, Stefani M, Ramponi G, Dobson CM (2002) Kinetic partitioning of protein folding and aggregation. Nat Struct Biol 9(2):137–143
26. Ventura S (2005) Sequence determinants of protein aggregation: tools to increase protein solubility. Microb Cell Fact 4(1):11
27. Steward A, Adhya S, Clarke J (2002) Sequence conservation in Ig-like domains: the role of highly conserved proline residues in the fibronectin type III superfamily. J Mol Biol 318(4):935–940
28. Chiti F, Calamai M, Taddei N, Stefani M, Ramponi G, Dobson CM (2002) Studies of the aggregation of mutant proteins in vitro provide insights into the genetics of amyloid diseases. Proc Natl Acad Sci U S A 99(Suppl 4):16419–16426

29. Schwartz R, King J (2006) Frequencies of hydrophobic and hydrophilic runs and alternations in proteins of known structure. Protein Sci 15(1):102–112

30. West MW, Wang W, Patterson J, Mancias JD, Beasley JR, Hecht MH (1999) De novo amyloid proteins from designed combinatorial libraries. Proc Natl Acad Sci U S A 96:11211–11216

31. Broome BM, Hecht MH (2000) Nature disfavors sequences of alternating polar and non-polar amino acids: implications for amyloidogenesis. J Mol Biol 296(4):961–968

32. Petkova AT, Leapman RD, Guo Z, Yau WM, Mattson MP, Tycko R (2005) Self-propagating, molecular-level polymorphism in Alzheimer's beta-amyloid fibrils. Science 307(5707):262-265

33. Nelson R, Eisenberg D (2006) Recent atomic models of amyloid fibril structure. Curr Opin Struct Biol 16(2):260–265

34. DuBay KF, Pawar AP, Chiti F, Zurdo J, Dobson CM, Vendruscolo M (2004) Prediction of the absolute aggregation rates of amyloidogenic polypeptide chains. J Mol Biol 341(5):1317–1326

35. Morel B, Varela L, Azuaga AI, Conejero-Lara F (2010) Environmental conditions affect the kinetics of nucleation of amyloid fibrils and determine their morphology. Biophys J 99(11):3801–3810

36. Ventura S, Zurdo J, Narayanan S, Parreno M, Mangues R, Reif B, Chiti F, Giannoni E, Dobson CM, Aviles FX, Serrano L (2004) Short amino acid stretches can mediate amyloid formation in globular proteins: the Src homology 3 (SH3) case. Proc Natl Acad Sci U S A 101(19):7258-7263

37. Rousseau F, Serrano L, Schymkowitz JW (2006) How evolutionary pressure against protein aggregation shaped chaperone specificity. J Mol Biol 355(5):1037–1047

38. Ritter C, Maddelein ML, Siemer AB, Luhrs T, Ernst M, Meier BH, Saupe SJ, Riek R (2005) Correlation of structural elements and infectivity of the HET-s prion. Nature 435(7043):844-848

39. Linding R, Schymkowitz J, Rousseau F, Diella F, Serrano L (2004) A comparative study of the relationship between protein structure and beta-aggregation in globular and intrinsically disordered proteins. J Mol Biol 342:345–353

40. Pechmann S, Levy ED, Tartaglia GG, Vendruscolo M (2009) Physicochemical principles that regulate the competition between functional and dysfunctional association of proteins. Proc Natl Acad Sci U S A 106(25):10159–10164

41. Castillo V, Ventura S (2009) Amyloidogenic regions and interaction surfaces overlap in globular proteins related to conformational diseases. PLoS Comput Biol 5(8):e1000476

42. Cremades N, Cohen SI, Deas E, Abramov AY, Chen AY, Orte A, Sandal M, Clarke RW, Dunne P, Aprile FA, Bertoncini CW, Wood NW, Knowles TP, Dobson CM, Klenerman D (2012) Direct observation of the interconversion of normal and toxic forms of alpha-synuclein. Cell 149(5):1048–1059

43. Fandrich M, Forge V, Buder K, Kittler M, Dobson CM, Diekmann S (2003) Myoglobin forms amyloid fibrils by association of unfolded polypeptide segments. Proc Natl Acad Sci U S A 100(26):15463–15468

44. Chiti F, Dobson CM (2009) Amyloid formation by globular proteins under native conditions. Nat Chem Biol 5(1):15–22

45. Plakoutsi G, Taddei N, Stefani M, Chiti F (2004) Aggregation of the Acylphosphatase from Sulfolobus solfataricus: the folded and partially unfolded states can both be precursors for amyloid formation. J Biol Chem 279(14):14111–14119

46. Bemporad F, Vannocci T, Varela L, Azuaga AI, Chiti F (2008) A model for the aggregation of the acylphosphatase from Sulfolobus solfataricus in its native-like state. Biochim Biophys Acta 1784(12):1986–1996

47. Sawaya MR, Sambashivan S, Nelson R, Ivanova MI, Sievers SA, Apostol MI, Thompson MJ, Balbirnie M, Wiltzius JJ, McFarlane HT, Madsen AO, Riekel C, Eisenberg D (2007) Atomic structures of amyloid cross-beta spines reveal varied steric zippers. Nature 447(7143):453–457

48. Elam JS, Taylor AB, Strange R, Antonyuk S, Doucette PA, Rodriguez JA, Hasnain SS, Hayward LJ, Valentine JS, Yeates TO, Hart PJ (2003) Amyloid-like filaments and water-filled nanotubes formed by SOD1 mutant proteins linked to familial ALS. Nat Struct Biol 10(6):461-467

49. Olofsson A, Ippel JH, Wijmenga SS, Lundgren E, Ohman A (2004) Probing solvent accessibility of transthyretin amyloid by solution NMR spectroscopy. J Biol Chem 279(7):5699–5707
50. Janowski R, Kozak M, Abrahamson M, Grubb A, Jaskolski M (2005) 3D domain-swapped human cystatin C with amyloidlike intermolecular beta-sheets. Proteins 61(3):570–578
51. Sambashivan S, Liu Y, Sawaya MR, Gingery M, Eisenberg D (2005) Amyloid-like fibrils of ribonuclease A with three-dimensional domain-swapped and native-like structure. Nature 437(7056):266–269
52. Vendruscolo M, Tartaglia GG (2008) Towards quantitative predictions in cell biology using chemical properties of proteins. Mol BioSyst 4(12):1170–1175
53. Castillo V, Grana-Montes R, Sabate R, Ventura S (2011) Prediction of the aggregation propensity of proteins from the primary sequence: aggregation properties of proteomes. Biotechnol J 6(6):674–685
54. de Groot NS, Castillo V, Grana-Montes R, Ventura S (2012) AGGRESCAN: method, application, and perspectives for drug design. Methods Mol Biol 819:199–220
55. Garbuzynskiy SO, Lobanov MY, Galzitskaya OV (2010) FoldAmyloid: a method of prediction of amyloidogenic regions from protein sequence. Bioinformatics 26(3):326–332
56. Trovato A, Seno F, Tosatto SC (2007) The PASTA server for protein aggregation prediction. Protein Eng Des Sel 20(10):521–523
57. Tartaglia GG, Vendruscolo M (2008) The Zyggregator method for predicting protein aggregation propensities. Chem Soc Rev 37(7):1395–1401
58. Maurer-Stroh S, Debulpaep M, Kuemmerer N, Lopez de la Paz M, Martins IC, Reumers J, Morris KL, Copland A, Serpell L, Serrano L, Schymkowitz JW, Rousseau F (2010) Exploring the sequence determinants of amyloid structure using position-specific scoring matrices. Nat Methods 7(3):237–242
59. Thompson MJ, Sievers SA, Karanicolas J, Ivanova MI, Baker D, Eisenberg D (2006) The 3D profile method for identifying fibril-forming segments of proteins. Proc Natl Acad Sci U S A 103(11):4074–4078
60. Chennamsetty N, Voynov V, Kayser V, Helk B, Trout BL (2009) Design of therapeutic proteins with enhanced stability. Proc Natl Acad Sci U S A 106(29):11937–11942

Chapter 11
Importance of Protonation States for the Binding of Ligands to Pharmaceutical Targets

Alberto Podjarny and Eduardo Howard

Abstract Protonation states of protein residues in ligand binding sites determine the electrostatic potential, which is essential to understand the interactions of the ligand and the protein. The case of aldose reductase is shown as an example. Inhibitors bind to the active site and to the nearby selectivity pocket. The case of two inhibitors, IDD 594 and Fidarestat, is discussed. The binding properties are determined by the protonation states of the protein residues, notably of His 110, and by the protonation state of the ligand, which can change in the case of Fidarestat. In this latter case the change in the charged state of the ligand during binding, from neutral to negative, combines the advantage of strong potency (charged state) and favorable pharmacokinetics (neutral state).

11.1 Introduction

Understanding and predicting protein-ligand binding is essential for structure-based drug design [1]. In particular, the first stage of the process involves identifying a "hit", which is a ligand that binds to the target protein with significant affinity. The ligand binding process involves several factors, such as structural complementarity, charge interactions and desolvation effects. Therefore, structural rearrangements that accompany protein-ligand binding have been extensively explored [2–4], as well as structure-energy relationships during the binding process [5–7].

An important component of the binding energy is provided by electrostatic interactions [8], which are strongest between fully charged ligands and protein residues but are also significant between partially charged atoms. The charge state is

A. Podjarny (✉)
Department of Integrative Biology, IGBMC, CNRS-INSERM-UDS, 1 rue Laurent Fries, 67404 Illkirch, France
e-mail: podjarny@igbmc.fr

E. Howard
Department of Biophysics, IFLYSIB, CONICET- UNLP, 59 N° 789, La Plata, Argentina

© Springer Science+Business Media Dordrecht 2015
G. Scapin et al. (eds.), *Multifaceted Roles of Crystallography in Modern Drug Discovery*, NATO Science for Peace and Security Series A: Chemistry and Biology, DOI 10.1007/978-94-017-9719-1_11

dependent on the protonation state of the protein residues, which is not necessarily known a priori for several cases, like histidine, aspartic, glutamic and lysine [9]. It is therefore important to study in detail the protonation states of the protein-binding site and of the ligand, taking into account that it can eventually be modified during complex formation.

As the charge state can have a profound effect upon ligand binding, the question of whether changes in protein or ligand charge state occur during the process of ligand binding is important. However, not much is known about the modifications in the charge state of the protein or the ligand upon complex formation, which are due to changes in pK values and ionization states of residues in the protein or polar moieties in the ligand. Altering the charge state of the binding interface via specific mutations can affect protein-ligand binding affinity, as shown both experimentally [10, 11] and theoretically [12], and can even be used to design complexes with higher affinity [13, 14] and in computational methods of binding prediction. In particular, the docking accuracy [15] and binding affinity predictions [16] improve when the quantum-mechanical calculations are used for the redistribution of ligand charges upon binding. This effect has been confirmed in another study, in which the prediction of ionization states allowed the accurate calculation of binding affinities between HIV protease and some inhibitors [17].

A powerful technique to directly observe hydrogen (deuterium) atoms and there-fore protonation states is protein crystallography. However, its application has been limited so far as it is necessary to use either subatomic resolution X-Ray diffraction or neutron diffraction, both of which are difficult to implement. This situation is rapidly changing, as recent developments in neutron protein crystallography are improving this picture and opening the possibility of determining protonation states in more cases.

In this chapter, we will study in detail one case, Aldose Reductase, which diffracts to subatomic resolution (beyond 1.0 Å). We will describe the complexes with two ligands: IDD 594 [18], in which the protonation states and hydrogen orientation in the protein are critical for binding, and Fidarestat (SNK-860) [19], in which the protonation state of the protein and the ligand change during complex formation.

11.2 Aldose Reductase

Human Aldose Reductase (h-AR; EC 1.1.1.21) is a NADPH-dependent enzyme that reduces a wide range of substrates, such as aldehydes, aldoses and corticosteroids. As it reduces D-glucose into D-sorbitol, it is believed to cause severe degenerative complications of diabetes [20]. The catalytic reaction involves a hydride transfer from carbon C4 of the coenzyme NADPH, which becomes $NADP^+$, and a proton donation from the enzyme [21]. In order to visualize the h-AR interactions with NADPH cofactor, substrates, and inhibitors and to determine the details of the catalytic mechanism, several X-ray analysis [22–24], site-directed mutagenesis [25–28] and modeling studies [29] have been performed. The extraordinary quality of h-AR crystals, which diffract X-Rays to 0.66 Å resolution, allows structural studies at a level of detail not available before for an enzyme of this size.

The h-AR protein folds as a $(\beta/\alpha)_8$ TIM barrel with the nicotinamide ring of the NADP$^+$ buried at the bottom of a deep cleft [22] (Fig. 11.1). The catalytic carbon C4 of the nicotinamide ring of NADP$^+$ is accessible through this cleft, thus defining the catalytic zone of the active site.

Residues Asp43, Tyr48, Lys77, His110, Asn160 and Gln183, highly conserved in all enzymes of the aldo-ketoreductase superfamily, form a tight network of hydrogen bonds linked to the nicotinamide ring (Fig. 11.2).

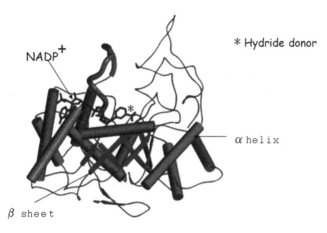

Fig. 11.1 Structure of human Aldose Reductase. *Cylinders*: α-helix. *Arrows*: β-sheets. *Ligand*: NADP$^+$

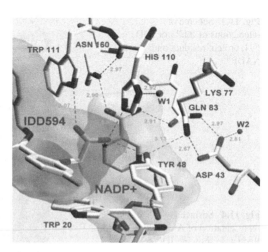

Fig. 11.2 H-bond network in the active site, showing the interactions between residues and the inhibitor IDD 594. Shadowed cavity: Active site

11.2.1 Aldose Reductase Complexes with IDD 594

Ligands bind inside the active site pocket. Some of them, like IDD 594, have a charged carboxylate head. This head makes a strong electrostatic interaction with NADP$^+$ (Fig. 11.3). In general, the positively charged NADP$^+$ in the active site pocket contributes to the "anionic binding site" where negatively charged ligands bind. Also, the tail of the ligand opens a new cavity, known as the "specificity pocket" (Figs. 11.3 and 11.4). IDD 594 is a very potent ligand (IC$_{50}$ = 30nM), but the fact that it is charged implies poor pharmacokinetic properties, as charged molecules are trapped in membrane barriers.

The main binding interaction of inhibitor IDD 594 is through the carboxylate head, which makes a strong electrostatic interaction with NADP$^+$ (Fig. 11.5a) and hydrogen bonds with three protein residues, Tyr 48, His 110 and Trp 111 (Fig. 11.5b). His 110 has two possible sites of protonation: Nδ1 and Nε2, leading to three possible states, two single ones and a double one. The protonation state of His 110 is important, as it determines the donor or acceptor capacity for hydrogen bonds. In the case of IDD 594, the subatomic resolution structure (0.66 Å) has clearly shown that His 110 is singly protonated (position Nε2, Fig. 11.5b), and it makes a short (2.67 Å) H-bond with the carboxylate head.

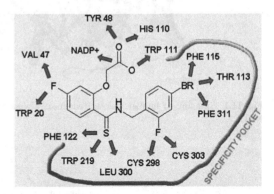

Fig. 11.3 Scheme of interactions of inhibitor IDD 594, protein residues and NADP$^+$

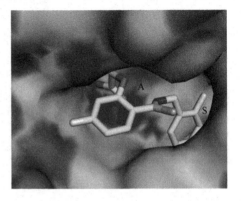

Fig. 11.4 Surface representation of Aldose Reductase showing IDD 594 binding in the active site (*A*) and the specificity pocket (*S*)

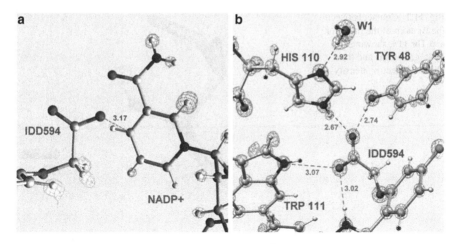

Fig. 11.5 (**a**) Electrostatic contact of IDD 594 and NADP$^+$. (**b**) H-bonds of IDD 594 with Tyr 48, His 110 and Trp 111. Two electron density maps are shown, $2F_o$-F_c (non-H atoms) and F_o-F_c (H-atoms)

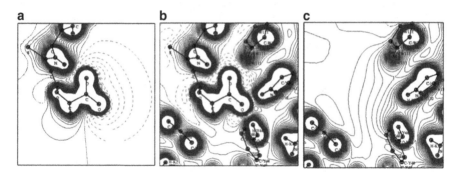

Fig. 11.6 Electrostatic potentials: (**a**) Inhibitor IDD 594 alone; (**b**) complex; (**c**) Inhibitor protein active site alone

This structure has been used for calculations of the electrostatic potential using the DFT technique [30]. The result of these calculations shows a clear complementarity between the negative electrostatic potential of the inhibitor alone (Fig. 11.6a, dashed lines) and the positive one of the protein active site (Fig. 11.6c, full lines) giving the protein-ligand complex (Fig. 11.6b).

Another important contact is between the Br atom of the inhibitor, which binds in the specificity pocket, and the residue Thr 113 (Fig. 11.7). In this case, the DFT calculations show a strong electrostatic interaction between a positive charge along the polar axis of the Br atom (Fig. 11.8a) and the negative charge associated with the Oγ atom of Thr 113 (Fig. 11.8c) which come together to form a short contact (Fig. 11.8b). Note that in this case the H atom bound to Oγ does not intervene, but it makes an internal H bond (O-O distance 2.81 Å) with the main chain oxygen of Thr 113 (Fig. 11.7).

Fig. 11.7 Contact between
the Br atom of the inhibitor
and Thr 113, showing the
hydrogen atoms and their
observed electron density

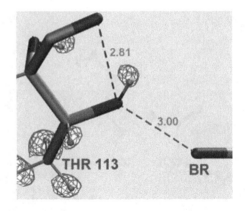

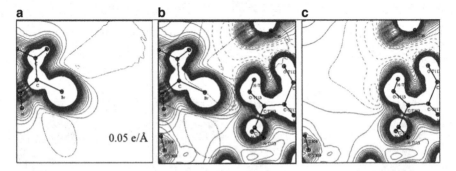

0.05 e/Å

Fig. 11.8 Electrostatic potential (**a**) Inhibitor IDD 594 alone; (**b**) Complex; (**c**) Inhibitor protein active site alone

11.2.2 Aldose Reductase Complexed with Fidarestat

Another ligand, Fidarestat (also known as SNK-860), has a hydantoin group (Fig. 11.9a), which is not charged in solution. The structure of Fidarestat complexed with h-AR has been solved at 0.92 Å resolution. The position of the inhibitor was clearly seen in the subatomic resolution structure (Fig. 11.9b) [19].

Fidarestat binds into the active site and does not open the specificity pocket (Fig. 11.10a). The binding at the active site implies the same residues than the carboxylate ligands, His 110, Tyr 148 and Trp 111 (Fig. 11.10b). This ligand has better pharmacokinetic properties than IDD 594 and good potency ($IC_{50} = 9$ nM).

Usually, very potent h-AR inhibitors are negatively charged, which is not the case of Fidarestat in solution. To explain its high potency, we studied the possibility that Fidarestat changes its charge state from neutral to negative when binding, employing the very high resolution structure as a tool to determine the protonation states. In particular, we looked at the hydrogen atom position in the H-bond between His 110, which was seen previously as protonated in the Nε2 position, and the hydantoin group, in which the N atom is protonated in solution.

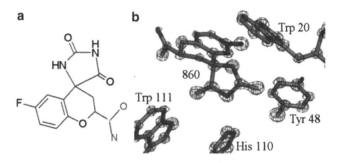

Fig. 11.9 (**a**) Formula of Fidarestat. (**b**) Electron density map showing the position of Fidarestat

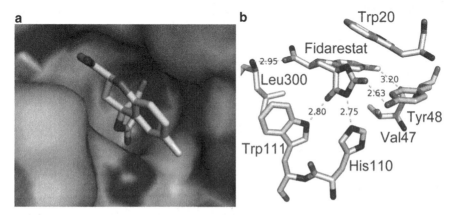

Fig. 11.10 Interactions of Fidarestat. (**a**) Surface representation showing Fidarestat binding in the active site cleft. (**b**) Contacts of Fidarestat with Tyr 48, His 110, Trp 111 and Leu 300

The difference map (Fig. 11.11a) shows actually two possible positions for the H atom, one linked to His 110 and one linked to Fidarestat. In the same structure, anomalous maps enabled to identify that the water molecule H-bonded to His 110 is replaced in 60 % of the cases by a Cl⁻ ion (Fig. 11.11b).

This information was used to propose a mechanism for the binding of Fidarestat, as follows (Fig. 11.12):

(I) In the native configuration, the active site is occupied by a citrate ion (observed in the native structure) that compensates the charge of $NADP^+$.

(II) Part of the inhibitor binds in neutral state, displacing the citrate ion. Simultaneously, a Cl⁻ ion replaces the water molecule. This happens for Fidarestat in 60 % of the cases (as indicated by the Cl⁻ occupation).

(III) In Fidarestat, the presence of the Cl⁻ ion shifts positively the pKa of His110. The histidine accepts the proton from the N1 atom of the hydantoin. An electrostatic interaction appears between the charged histidine and the charged inhibitor.

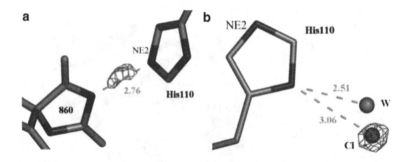

Fig. 11.11 (a) Contact between His 110 and Fidarestat showing the elongated electron density of the H-atom corresponding to two possible positions. (b) Water molecule and Cl⁻ ion linked to His 110. The anomalous map showing the Cl⁻ position is shown

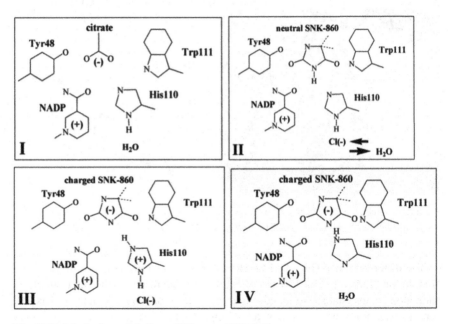

Fig. 11.12 Mechanism of binding of Fidarestat (SNK-860)

(IV) Alternatively, a charged population of the inhibitor is bound in a way similar to that observed for the carboxylic acid inhibitors. In Fidarestat, the occupation of the water molecule bound to His110 suggests that this happens in 40 % of the cases.

This mechanism shows how the inhibitor can bind in a neutral form and become charged in the binding site, therefore fulfilling both the conditions of good pharmacokinetics and high potency.

11.3 Conclusions

The Aldose Reductase case has been used as an example of the use of subatomic resolution crystallography to determine protonation states in two cases:

1. For the complex with IDD 594, the charge complementarity depends on the protonation state of His 110 and of the orientation of the hydroxyl hydrogen of Thr 113.
2. For the complex with Fidarestat, the favorable properties of the inhibitor are due to its ability to change its charge state when binding, through an interaction with His 110. It can therefore remain neutral in solution, improving the pharmacokinetic properties, and become charged when bound, improving the potency.

This illustrates the importance of protonation states and charge complementarity for obtaining very potent inhibitors, which can be an essential tool in rational drug design.

Acknowledgments We thank Andre Mitschler and Alexandra Cousido-Siah for their efforts in crystallization and data collection, the SBC staff for their support in data collection; N. Muzet and B. Guillot for electrostatic calculations, and Ossama El-Kabbani for his participation in the Fidarestat project. This work was supported by the Centre National de la Recherché Scientifique (CNRS), by collaborative projects CNRS-CONICET, CNRS-CERC and CNRS-NSF (INT-9815595), by Ecos Sud, by USA Federal funds from the National Cancer Institute (Contract No. NO1-CO-12400) and the National Institutes of Health, by the Institut National de la Santé et de la Recherché Médicale and the Hôpital Universitaire de Strasbourg (H.U.S), and by the Institute for Diabetes Discovery, Inc. through a contract with the CNRS, and in part by the U.S. Department of Energy, Office of Biological and Environmental Research under contract No. W-31-109-ENG-38. EH is a member of Carrera del Investigador, CONICET, Argentina.

References

1. Jorgensen WL (2004) The many roles of computation in drug discovery. Science 303(5665):1813–1818
2. Goh CS, Milburn D, Gerstein M (2004) Conformational changes associated with protein-protein interactions. Curr Opin Struct Biol 14(1):104–109
3. Holmes KC, Angert I, Kull FJ, Jahn W, Schroder RR (2003) Electron cryo-microscopy shows how strong binding of myosin to actin releases nucleotide. Nature 425(6956):423–427
4. Boehr DD, Wright PE (2008) Biochemistry. How do proteins interact? Science 320(5882):1429–1430
5. Sheinerman FB, Honig B (2002) On the role of electrostatic interactions in the design of protein-protein interfaces. J Mol Biol 318(1):161–177
6. Bordner AJ, Abagyan R (2005) Statistical analysis and prediction of protein-protein interfaces. Proteins 60(3):353–366
7. Ofran Y, Rost B (2003) Predicted protein-protein interaction sites from local sequence information. FEBS Lett 544(1–3):236–239
8. Kellenberger E, Dejaegere A (2011) Molecular modelling of ligand-macromolecule complexes. In: Podjarny A, Dejaegere A, Kieffer B (eds) Biophyiscal approaches determining ligand binding to macromolecular targets. Royal Society of Chemistry, Cambridge, pp 300–356

9. Petukh M, Stefl S, Alexov E (2013) The role of protonation states in ligand-receptor recognition and binding. Curr Pharm Des 19(23):4182–4190
10. Clackson T, Wells JA (1995) A hot spot of binding energy in a hormone-receptor interface. Science 267(5196):383–386
11. Lowman HB, Cunningham BC, Wells JA (1991) Mutational analysis and protein engineering of receptor-binding determinants in human placental lactogen. J Biol Chem 266(17): 10982–10988
12. Massova I, Kollman PA (1999) Computational alanine scanning to probe protein-protein interactions: a novel approach to evaluate binding free energies. J Am Chem Soc 121: 8133–8143
13. Kangas E, Tidor B (1999) Charge optimization leads to favorable electrostatic binding free energy. Phys Rev E Stat Phys Plasmas Fluids Relat Interdiscip Topics 59(5 Pt B):5958–5961
14. Green DF, Tidor B (2005) Design of improved protein inhibitors of HIV-1 cell entry: optimization of electrostatic interactions at the binding interface. Proteins 60(4):644–657
15. Cho AE, Guallar V, Berne BJ, Friesner R (2005) Importance of accurate charges in molecular docking: quantum mechanical/molecular mechanical (QM/MM) approach. J Comput Chem 26(9):915–931
16. Donnini S, Villa A, Groenhof G, Mark AE, Wierenga RK, Juffer AH (2009) Inclusion of ionization states of ligands in affinity calculations. Proteins 76(1):138–150
17. Wittayanarakul K, Hannongbua S, Feig M (2008) Accurate prediction of protonation state as a prerequisite for reliable MM-PB(GB)SA binding free energy calculations of HIV-1 protease inhibitors. J Comput Chem 29(5):673–685
18. Howard EI, Sanishvili R, Cachau RE, Mitschler A, Chevrier B, Barth P, Lamour V, Van Zandt M, Sibley E, Bon C, Moras D, Schneider TR, Joachimiak A, Podjarny AD (2004) Ultra-high resolution drug design I: human aldose reductase – inhibitor complex at 0.66 Å shows experimentally protonation states and atomic interactions which have implications for the inhibition mechanism. Proteins Struct Funct Genet 55:792–804
19. El-Kabbani O, Darmanin C, Schneider TR, Hazemann I, Ruiz F, Oka M, Joachimiak A, Schulze-Briese C, Tomizaki T, Mitschler A, Podjarny A (2004) Ultrahigh resolution drug design. II. Atomic resolution structures of human aldose reductase holoenzyme complexed with Fidarestat and Minalrestat: implications for the binding of cyclic imide inhibitors. Proteins 55(4):805–813
20. Yabe-Nishimura C (1998) Aldose reductase in glucose toxicity: a potential target for the prevention of diabetic complications. Pharmacol Rev 50:21–33
21. Wermuth B (1985) Enzymology of carbonyl metabolism 2: aldehyde dehydrogenase, aldo-keto reductase, and alcohol dehydrogenase. Alan R. Liss Inc., New York
22. Rondeau JM, Tete-Favier F, Podjarny A, Reymann JM, Barth P, Biellmann JF, Moras D (1992) Novel NADPH-binding domain revealed by the crystal structure of aldose reductase. Nature 355:469–472
23. Wilson DK, Bohren KM, Gabbay KH, Quiocho FA (1992) An unlikely sugar substrate site in the 1.65 A structure of the human aldose reductase holoenzyme implicated in diabetic complications. Science 257(5066):81–84
24. Tete-Favier F, Barth P, Mitschler A, Podjarny A, Rondeau J-M, Urzhumtsev A, Biellmann J-F, Moras D (1995) Aldose reductase from pig lens. Eur J Med Chem 30S(30):589s–603s
25. Tarle I, Borhani DW, Wilson DK, Quiocho FA, Petrash JM (1993) Probing the active site of human aldose reductase. Site-directed mutagenesis of Asp-43, Tyr-48, Lys-77, and His-110. J Biol Chem 268:25687–25693
26. Schlegel BP, Jez JM, Penning TM (1998) Mutagenesis of 3 alpha-hydroxysteroid dehydrogenase reveals a "push-pull" mechanism for proton transfer in aldo-keto reductases. Biochemistry 37:3538–3548
27. Schlegel BP, Ratnam K, Penning TM (1998) Retention of NADPH-linked quinone reductase activity in an aldo-keto reductase following mutation of the catalytic tyrosine. Biochemistry 37:11003–11011

28. Bohren KM, Grimshaw CE, Lai CJ, Harrison DH, Ringe D, Petsko GA, Gabbay KH (1994) Tyrosine-48 is the proton donor and histidine-110 directs substrate stereochemical selectivity in the reduction reaction of human aldose reductase: enzyme kinetics and crystal structure of the Y48H mutant enzyme. Biochemistry 33:2021–2032
29. Lee YS, Chen Z, Kador PF (1998) Molecular modeling studies of the binding modes of aldose reductase inhibitors at the active site of human aldose reductase. Bioorg Med Chem 6: 1811–1819
30. Muzet N, Guillot B, Jelsch C, Howard E, Lecomte C (2003) Electrostatic complementarity in an aldose reductase complex from ultra-high-resolution crystallography and first-principles calculations. Proc Natl Acad Sci U S A 100(15):8742–8747

Chapter 12
Protein-Protein Interactions: Structures and Druggability

David B. Ascher, Harry C. Jubb, Douglas E.V. Pires, Takashi Ochi, Alicia Higueruelo, and Tom L. Blundell

Abstract While protein-protein interfaces have promised a range of benefits over conventional sites in drug discovery, they present unique challenges. Here we describe recent developments that facilitate many aspects of the drug discovery process – including characterization and classification of interfaces, identifying druggable sites and strategies for inhibitor development.

12.1 Historical Background

Over the past 40 years structure-guided approaches have become increasingly central to the discovery and design of new therapeutics. Initially the focus was on modification of natural substrates or molecules known to bind and inhibit enzymes or cell surface receptors. Over the past 20 years new hits have been derived largely from screening using either whole cell assays or enzyme assays with chemical libraries that may number hundreds of thousands of drug-like compounds. Chemical libraries have been refined to make them more closely compliant with the Lipinsky Rule-of-Five that requires molecules to be less large (MW <500), less lipophilic (LogP <5), less flexible and have the requisite number of hydrogen bond donors and acceptors, features of molecules that have led to successful therapeutics [1]. This has led to many successful drugs reaching the market but at exponentially increasing costs.

One feature of drug-discovery campaigns has been the tendency to select targets that have been defined as "druggable", often leading to a focus on large enzyme superfamilies where one member has been target for a successful drug campaign; examples include aspartic proteinases [2–4], metallo proteinases [5–9], transferases [10, 11] and protein kinases, all of which have well defined concave active sites [12]. Designs have often been mechanism based, reflecting either the co-factor or

D.B. Ascher • H.C. Jubb • D.E.V. Pires • T. Ochi • A. Higueruelo • T.L. Blundell (✉)
Department of Biochemistry, Sanger Building, University of Cambridge, Tennis Court Road, Cambridge CB2 1GA, UK
e-mail: dascher@svi.edu.au; tom@cryst.bioc.cam.ac.uk

© Springer Science+Business Media Dordrecht 2015
G. Scapin et al. (eds.), *Multifaceted Roles of Crystallography in Modern Drug Discovery*, NATO Science for Peace and Security Series A: Chemistry and Biology, DOI 10.1007/978-94-017-9719-1_12

an enzyme intermediate or transition state. These have been optimized to useful leads often using structure-guided approaches. More recently with structures of membrane proteins available there has interest has revived in the GPCRs and other members of large membrane protein superfamilies that have been classically successful targets for phenotypic screening approaches.

The development of fragment-based drug discovery has allowed the more effective exploration of chemical space for ligands targeting a particular binding site by using much smaller, chemically diverse libraries composed of smaller molecules, thereby decreasing complexity. These small molecules, usually between 100 and 300 Da and consistent with the Rule-of-Three [13], have been termed 'fragments'. Detection and development of these fragments are complicated by the fact that they typically bind with much weaker affinities, in the mM range, than larger molecules. Wells and colleagues pioneered a way around this problem, tethering them by exploiting thiol-containing fragments that react through a reversible disulphide bond with a cysteine residue that has been engineered into a protein [14]. This proved particularly successful for protein-protein interactions or where sites require trapping a particular conformer. Another approach that has been successfully used to screen and validate fragment binding has relied on highly sensitive biophysical methods [15, 16]; including nuclear magnetic resonance (NMR; [17–19]), X-ray crystallography [20, 21], surface plasmon resonance (SPR; [22–24]), differential scanning fluorimetry (DSF; [25–27]) or isothermal calorimetry (ITC; [28–31]). Fragment hits identified from either approach are subsequently elaborated by crosslinking or 'growing' the fragment, while maintaining strong interactions for each group added. This method has proven very effective when used as part of a structure-guided approach [18, 32, 33].

12.2   Obtaining Selectivity with Multiprotein Systems

It has become increasingly evident over the past decade that it is difficult to obtain selectivity, especially with transition and intermediate state analogues of enzymes or those targeting co-factor binding sites. In particular the challenges with targeting protein kinases has become particularly clear as pharmaceutical companies have increased the numbers of superfamily members that can be assayed. Much of the optimism of getting very good selectivity with protein kinase inhibitors by exploiting sub-pockets around the ATP binding site has been moderated by the discovery that sub-clusters of protein kinases with similar co-factor binding sites are recognised by many of the molecules previously thought to be selective.

One of the ways of improving selectivity is to move away from targeting active sites towards regulatory multiprotein systems that are critical to cell activity [34]. A particularly fruitful line of investigation has been to target the various protein interactions that regulate different members of one superfamily. Thus, receptor tyrosyl kinases have multiple different regulatory extracellular regions that may interact with secondary receptors as well as their ligands, often leading to clustering on the cell surface. For example fibroblast growth factor receptor (FGFR) recognizes the FGF ligand at a binding site between extracellular domains d2 and d3 but a secondary receptor the proteoglycan heparan sulphate is also critical for activity and tends to mediate clustering of receptor and ligand [35–38]. The MET receptor tyrosyl kinase differs in its extracellular region that recognizes a very different protein growth factor, HGF/SF [39–41]; in this case the heparan sulphate secondary receptor is important for some splice forms. Other receptor tyrosyl kinases such as epidermal growth factor (EGF) receptor and the insulin receptor are related evolutionarily but have different domain organizations of the region recognizing the ligands. These systems give opportunities for greater selectivity of inhibitors, either by targeting protein-protein interfaces in the assemblies directly (orthosteric inhibitors; see ref [42] for review) or indirectly through allosteric binding sites that stabilize conformers incommensurate with ligand binding at another site, as for example identified for the FGF receptor [43].

Intracellular signaling pathways are also regulated by multiprotein systems of similar complexity. The model of receptor activation that leads to a pathway of interactions, somewhat resembling the metabolic pathways familiar to biochemists, appears to be giving way to the idea that large multiprotein systems often assemble to regulate many intracellular kinases and these mediate interactions between the cell membrane and various cytoplasmic and nuclear targets [44]. In the nucleus complex regulatory systems, for example mediating DNA double-strand break repair, involve multi-component systems. Non-homologous end-joining (NHEJ) requires many factors: the Ku70 and 80 heterodimer that assembles on double-strand breaks (DSBs); the scaffolding proteins XRCC4 and XLF which interact with Ku; the key protein for recruiting NHEJ proteins at DNA ends, the ligase (DNA ligase IV) that joins the ends [45]; and the DNA-PKcs that is involved in signalling and regulating DNA repair. These proteins assemble together with other proteins such as the nuclease Artemis that also has binding sites for the ligase and DNA- PKcs [46]. Ku interacts directly with DNA-PKcs [47, 48], DNA ligase IV [49] and XLF [50] in a DNA-dependent manner, and recruits NHEJ proteins *in vivo* only when DSBs are generated.

Such complex multiprotein assemblies at the membrane, in the cytoplasm and in the nucleus often regulate cellular processes through co-location of various critical components. However, they also likely play a role in increasing signal to noise. Although binary interactions between two proteins would often occur opportunistically in the cell, especially in the cell membrane or in the limited environment of the nucleus or cytoplasm, a weak binary interaction followed by interactions of further components would give a cooperative but reversible assembly of a large multiprotein complex, allowing selective signaling regulation in the cell [44].

There are some occasions where binary systems are required in signaling and regulatory processes. These are often mediated by concerted folding and binding of one protein. This was recognized more than 30 years ago in the polypeptide hormones like glucagon which are disordered in solution but can associate with the receptor in a cooperatively formed secondary or supersecondary structure, often first binding an anchor residue, which forms a hotspot of the interaction [51]. Such concerted-folding-and-binding is found widely in intracellular systems. A good example is BRCA2 BRC4 repeat interaction with RAD51, in which a phenylalanine, the anchor residue, recognizes a well-defined pocket on the RAD51 [52]. This then also allows a much smaller pocket, a second hotspot to be recognized, the interaction probably being driven by unhappy water, leading to high selectivity for alanine. The remaining part of the BRC4 repeat then folds onto the surface of the RAD51 through a weaker and less well defined interaction involving folding and binding of a helix onto a further hotspot. The cooperative folding and binding constitutes provides a second mechanism for obtaining selectivity and has been widely studied for intracellular systems by Wright and coworkers [53].

The two mechanisms of gaining selectivity – co-operative assembly of multi protein regulatory assemblies or co-operative folding and binding – both involve protein-protein interactions. Here we review these interactions, focusing on how they are defined experimentally, the nature of the interfaces that mediate the interactions, the effects of mutations at protein-protein interfaces and their roles in genetic diseases, and the druggability of either the isolated interfaces or the interfaces themselves.

12.3 How to Define Structures of Multiprotein Assemblies

The complex nature of many of the regulatory assemblies demands different techniques to characterize the stoichiometry of the interactions that vary in space and time. These range quite broadly in the resolution and detail that they provide: from assessments of stoichiometries and molecular radii, to overall topology of a complex provided by small angle X-ray scattering and electron microscopy, to the atomic resolution provided by X-ray crystallography and increasingly single particle cryo-electron microscopy (Fig. 12.1a). The information obtained from these diverse techniques is often complementary and help provide an overall understanding of a given complex.

Nanospray mass spectrometry (MS) can accurately determine stoichiometry of macromolecular complexes as large as a few MDa [54]. An advantage of the technique is that samples do not need to be homogenous, and it can detect different oligomeric states existing in equilibrium together. Hence, it is very useful for macro-molecules that form dynamic complexes. Furthermore, nanospray MS can provide topological information of macromolecular assembles [55]. One of limitations is the buffer, which should be volatile, for example ammonium acetate. If nonvolatile chemicals are essential for the proteins or complex, alternative methods to study stoichiometry may be required. These include analytical ultracentrifugation (AUC) and size-exclusion chromatography combined with multi-angle light scattering (SEC-MALS), techniques that have been vital in establishing the oligomeric states of proteins previously thought to be monomeric, and the stoichiometry of their modulation by small molecules [56–58].

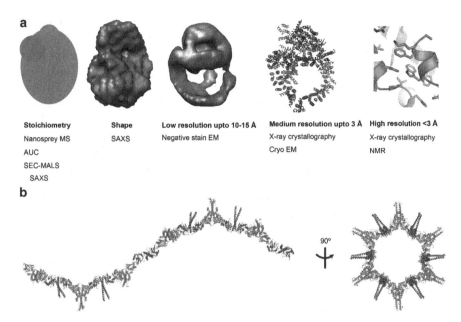

Fig. 12.1 Complementary-biophysical analyses provide different resolutions of structural information. (**a**) Comparison of biophysical techniques and resolutions. Structures of DNA-PKcs (BioIsis ID: 1DPKCY) [70], DNA-PK (EMDB ID: EMD-1210) [72], DNA-PKcs in complex with Ku80 C-terminus (PDB ID: 3KGV) [46], LigIV in complex with Artemis (PDB code ID: 3W1B) [85] were used to show examples of structural information obtained from SAXS, negative-stain EM and X-ray crystallography (low and high resolution). (**b**) Structure of the XRCC4/XLF filament. The structure of the XRCC4 (*dark grey*)/XLF (*light grey*) complex shows an alternative left-handed filament from two different views

Small angle X-ray scattering (SAXS) can also be used to determine stoichiometry of target proteins. A recent advance in SAXS allowed us to measure approximate molecular weights of only the proteins of interest from scattering data [59]. SAXS is also a good way to get a general idea of the solution structure and is particularly useful if structures of individual protein components are available [60, 61]. It can also be used to study protein complexes that have dynamic protein-protein interactions by combining with size-exclusion chromatography [62]. X-ray crystallography has proved the mainstay for defining complexes although interactions with smaller peptide ligands are often also accessible through NMR. Increasingly high-resolution cryoelectron microscopy is becoming very powerful and can even recognize small ligands binding to complex structures [63]. Isothermal calorimetry can be used to measure the thermodynamics of the interactions and surface plasmon resonance the kinetics.

For an example of the characterization of increasingly complex structures let us return to NHEJ, involved in repair of DNA-double-strand breaks and introduced above. The kinase catalytic subunit, DNA-PKcs, is a 460 kDa protein containing a long HEAT-repeat region of about 3,000 of the total 4,000 amino acid residues [64].

The C-terminal kinase region has α-helical FAT and FATC domains, in addition to the kinase domain. The high-resolution structure of the kinase region of mTOR, which is a paralog of DNA-PKcs, has a similar structure and was defined at high resolution X-ray crystallography [65]. The structure of the DNA-PK complex, which includes the Ku hetero-dimer, has been studied using EM, SAXS and X-ray crystallography. Early EM structures of DNA-PKcs identified two regions: crown/head and base/palm [66–68]. The resolution of a cryo-EM structure of DNA-PKcs was extended to 7 Å resolution many years later, revealing the secondary structure of the molecule [69]; however, the crystal structure of DNA-PKcs was required to determine unequivocally the location of the kinase [46]. The crystal structure of DNA-PKcs in complex with the C-terminal of Ku80 at 6.6 Å resolution showed that the HEAT repeats form a circular structure, consistent with the early EM models of DNA-PKcs. Recent EM and SAXS studies of DNA-PKcs showed that it undergoes a large conformational change upon autophosphorylation [70, 71], which is difficult to study using crystallography. Although the structure of whole DNA-PK complex has been investigated using EM and SAXS [70, 72], it remains unresolved where Ku70/80 binds on DNA-PKcs but both techniques consistently showed a hetero-hexameric complex of these proteins at two DNA ends. DNA-PK on DNA ends was also observed by atomic force microscopy (AFM) [73, 74]. AFM is particularly useful to see where proteins are bound on DNA because it can visualize naked DNA. Since these techniques are complementary to each other, accumulated structural studies of DNA-PK provide insights into how the complex binds DNA spatially.

In addition to DNA-PK, the structure of the ligase holoenzyme has been studied extensively and it provides a good example as to how the techniques complement each other. The DNA ligase IV, XRCC4 and XLF complex contributes to the last step of NHEJ [75]. Crystal structures of complexes of XRCC4 and XLF, structural paralogs that are both homodimers, show that they form a left-handed helical filament (Fig. 12.1b) [76–79]. A concentration-dependent formation of the filament is shown by gel filtration, nanospray mass spectrometry and SAXS [79] and a scanning force microscopy of the complex demonstrates that DNA stabilizes filament formation [76]. Interestingly, filament formation is inhibited by the strong and stable LigIV interaction with XRCC4 [80], likely due to the catalytic domains of LigIV, because LigIV/XRCC4 without the domains still forms the filament with XLF. SAXS studies of LigIV/XRCC4 indicate that the catalytic domains are flexibly tethered to a tandem repeat of BRCT domains [75, 81], which interact with XRCC4 [82, 83]. Thus, the dynamic nature of the catalytic domains prevents XLF from interacting with XRCC4. Negative-stain EM of LigIV/XRCC4 shows that the catalytic region is fixed at the N-terminal head domain of XRCC4 [84].

DNA-PK and ligase holoenzymes together with other NHEJ proteins such as Artemis are present at DNA ends. However, the dynamic nature of these protein-protein interactions makes it difficult to study structurally, which has high demand on homogeneity. Since we have atomic structures of the individual core NHEJ proteins, cryo-EM may be a reasonable technique to observe the entire complex. A key point for the success will depend upon how much we can stabilize the complex. For this optimization, nanospray MS and other techniques will help inform distribution of stoichiometry and stability of the complex.

These challenges in defining the spatial and temporal interactions of the NHEJ system are likely to be common to many of the multicomponent systems in the cell. Nevertheless they provide a basis for understanding their roles in cell regulation and signaling, and some data that is proving useful in the design of chemical tools that can be used to selectively modulate cell activity and provide the first clues about how to proceed in discovering candidate drugs.

12.4 Organization of PPI Information: Description of Piccolo and Credo

The wealth of data publicly available in the PDB [86] allows structural comparison of interacting proteins with a complete range of partners (solvents, small molecules, small peptides, saccharides, nucleic acids and other proteins). However, in order to use this information meaningfully and efficiently for drug discovery, the data in the PDB, a flat-file-based databank, need to be better organized. Efforts to do this range from resources specialized in one type of structural interaction, like beta-sheet or alpha-helix motifs [87, 88] or domain-domain interactions [89] to resources that emphasize the mechanistic aspects of interactions, like the ASD Allosteric Database [90, 91].

However, integrating specialized resources is challenging, as we discovered with the sister databases developed in our laboratory, describing structural interactions with atomic detail for protein-protein (PICCOLO, [92]), protein-nucleic acid (BIPA, [93]) and protein-ligand complexes (CREDO, [94]). A new CREDO database [95] has now been developed with the aim of enclosing under a single resource not only all pairwise atomic interactions of inter- and intra- molecular complexes from the PDB, but also disparate data relevant to drug discovery; these include SNP databases (OMIM and COSMIC [96, 97]), mappings to sequence data from UniProt [98] and EnsEMBL [99], ChEMBL [100] binding data and the small molecules fragmented with an enhanced RECAP [101] algorithm. In addition to providing relational data structures for storing protein structure data at model, chain, residue, ligand and atom level, CREDO provides chemoinformatic routines to analyze small molecule data, such as fingerprint generation, similarity and substructure searching and chemical fragmentation. Where entities in PDB structures are involved in non-bonding interactions, such as in protein-ligand and protein-protein interactions, pairwise atomic contacts are explicitly characterized, for example as hydrogen bonding, ion pair, metal complex or specific aromatic ring interactions. These pairwise atomic contacts are stored as structural interaction fingerprint SIFTs [102], used for clustering interactions to identify common patterns or to study molecular recognition, so making CREDO a comprehensive analysis platform for drug discovery. The information on intermolecular interactions is integrated with further chemical and biological data. The database implements useful data structures and algorithms such as cheminformatics routines to create a comprehensive analysis platform for drug discovery. The database can be accessed through a web-based interface, downloads of data sets and web services at http://structure.bioc.cam.ac. uk/credo.

The information in CREDO allows the user to move from target to target using a residue map (that links sequence to structure) to UniProt [98], or to analyze the intra-residue network interactions at the protein interfaces for correlation with hotspots.

12.5 Distribution of Protein-Protein Interactions and Pocket Size

Organizing the wealth of publicly available structural protein-protein interaction (PPI) data has made it clear that PPI interfaces, the chemical surfaces through which proteins interact with each other, come in many different shapes and sizes [92]. Multiprotein complexes can assemble from globular protomers, interacting with partners to form homo-complexes of sequence-and-structure-identical protomers; between partners of different sequence to form hetero-complexes, or between globular partners and short peptides or even longer polypeptides that are often disordered before binding [103]. There are also examples of peptide-peptide associations [104]. The range of different domains and peptides involved in interactions provides diversity in PPI binding sites [103, 105, 106]. While PPI interfaces have historically been described as being "flat and featureless" the growing number of examples of orthosteric PPI modulators [107], which compete for the binding site of one protein to another, speaks to the fact that not all PPI associations are as featureless and un-amenable to chemical modulation as was sometimes thought.

Concavity is generally accepted as a feature of many protein-ligand interactions, where binding deep into a protein's surface may maximize the interaction area between a protein and a small-molecule ligand, and where ligand binding may be entropically favorable through the ejection of water molecules from the protein's solvation shell into bulk solvent [108, 109]. Computational analyses have shown that where PPI interfaces have been successfully modulated, surface concavity at the binding site usually exists not in the single, large volume cavities found in "traditional" drug targets, but rather in multiple small, geometrically clustered concavities [110]. Examples of these kinds of concavities at interfaces have been identified as being used as "complemented pockets" by protein partners [111] involving deeply bound single residues. The concept is related to the "hotspot" hypothesis that single, buried residues or clusters of residues contribute a large proportion of interface interaction energy [112, 113].

The similar "anchor" hypothesis states that energetically important, solvent-buried residues at the interface are involved in initial, fast lock-and-key type recognition, followed by a more gradual relaxation of peripheral residues by an induced fit mechanism to form the mature interface [114, 115]. Exploitation of these anchor sites has been used in the design of orthosteric inhibitors [116, 117]. Protein-peptide interfaces may be particularly amenable to orthosteric modulation over other associations, at least in part due to their tendency to consist of single interacting "segments" where a linear binding epitope contributes a large proportion of the binding free energy of the interface [118], and the conformation of the globular partner is typically fixed such that the surface presents concavity to a disordered binding partner which subsequently folds on binding [103, 111]. The extensive buried surface area no doubt contributes to the affinity and compensates for the loss of entropy on folding. However, even in larger, more globular interfaces, there is a tendency for a single linear epitope to contribute large proportions of the interface's interaction energy [119, 120].

Much effort both in academia and industry has focused on the modulation of pairwise PPIs through the development of competitive, orthosteric inhibitors of interface formation. Although many interactions are dimeric, the disruption of a dimer forming part of a larger complex is likely to disrupt the assembly of that complex. As discussed above many biological systems involve the coordination of multiple protomers through strong but reversible cooperative binding to achieve high signal-to-noise ratios in cellular processes. There are advantages to attempting to chemically modulate multiprotein systems by targeting pre-bound structures, to either stabilize the complex to have a therapeutic effect or to ablate the formation of a higher order complex. "Interfacial inhibitors", or stabilizers, binding juxtaposed to a PPI binding site have been explored as a potential PPI modulation strategy [121]; in part successes may stem from the existence of better defined binding sites in the periphery of an existing interface [122]. There is also increasing interest in allosteric strategies, where the binding site is distal to the interface, for disrupting multiple protein assembly.

Systematic analysis in our laboratory of over 9,000 pairwise, non-overlapping PPI interfaces, organized in our databases and filtered for structure quality, has indicated that protein-peptide interfaces make more extensive use of concavity than other kinds of interfaces, both on average and at their deepest. However, in spite of being flatter on average, a large proportion of globular-globular interfaces make use of small pockets of concavity, through deeply bound residues. The landscapes of PPI interface surfaces make more subtle use of concavity than traditional targets, therefore requiring innovative approaches for drug discovery.

12.6   Mutations & Interfaces: Their Role in Diseases

Mutations are a natural consequence of evolution, and understanding how they interact with their partners can yield insights into protein function, diseases and help guide a range of experimental efforts including protein engineering and drug development. The first efforts to understand the effects of mutations focused on their ability to alter protein folding and stability. The pioneering method SDM

[123, 124] used a statistical potential energy function, derived from environment-specific residue propensities in structural families [125], in order to predict the change in free energy of folding upon mutation for a given protein.

The leap in computational power provided by new architectures, together with the rapid increase in protein experimental and structural data generation, has created new opportunities for enhancing the existing approaches (for a review of available methods see [126]). This scenario led to the development of mCSM [127], a novel machine learning method which is proving to be significantly more accurate and scalable than previous approaches. mCSM uses the concept of graph-based signatures to represent the three-dimensional environment of a wild-type residue, which are then used to train highly accurate predictive models, capable of quantitatively assessing the effects of mutations. These signatures have been proposed previously and successfully adapted in a range of applications including: protein inter-residue analysis [128], protein automatic structural classification and function prediction [129] and receptor-based ligand prediction [130]. However all these methods present only a portion of the story necessary to understand fully the effects of mutations as they did not take into account the multitude of interactions vital for normal cellular function. In this context, tools for assessing the impact of mutations on protein-protein interfaces are necessary.

Some of the early approaches to predict the effect of mutations on the binding free energy of specific protein-protein complexes included energy-function based approaches [131–133] and more computationally intensive calculations [134–141]. These methods, however, focused on mutations to alanine, which will be discussed further below. While alanine scanning is extremely important experimentally, in order to understand the broad array of genetic variations, and mutations in diseases, a more challenging demand was to develop methods capable of predicting the effects of any mutation. To this end we developed mCSM-PPI (http://bleoberis.bioc.cam.ac.uk/mcsm/protein_protein) which employs the mCSM signatures used as evidence to train predictive models based on experimentally measured effects on protein-protein affinity from the SKEMPI database [142]. The method has shown to be effective, performing extremely well in comparison with other methods, achieving a Pearson's correlation coefficient of up to 0.8 (as shown in Table 12.1).

Other methods have been recently described including BeAtMuSiC [143], ZEMu [144] and those described by Li et al. [145] and Moal et al. [146]. Since these methods were developed concurrently, a comparative analysis of their relative performance is of general interest. As shown in Table 12.1, mCSM can outperform methods that employ a range of different techniques, some of them computationally intensive, such as Molecular Dynamics and Molecular Mechanics.

Structure-based methods like mCSM-PPI are essential tools for understanding the relation between the quantitative effects of mutations in protein-protein affinity and their roles in Mendelian diseases and in cancer, as well as to shed light on the understanding of their mechanism of action.

One example of the usefulness of such methods is the recently published work of Gossage et al. [147] of mutations on von Hippel-Lindau disease (VHL) and their relation with propensity or risk of developing renal cell carcinoma (RCC).

Table 12.1 Predicting the effects of mutations on protein-protein interactions

Method	Technique	Data set (# of mutations)	Correlation (SE)	mCSM-PPI performance on similar data set	Web server	Refs.
mCSM-PPI	Structural signatures – ML	SKEMPI (2007)	0.80 (1.25)	N/A	Yes	[127]
BeAtMuSiC	Statistical potentials	SKEMPI (2007)	0.40 (1.80)	0.58 (1.55)	Yes	[143]
FoldX	Energy function	SKEMPI (1844)	0.37 (2.14)	0.58 (1.55)	Yes	[131]
Li et al.	MD	SKEMPI (1844)	0.58 (1.55)	0.58 (1.55)	No	[145]
ZEMu	MD	SKEMPI (1254)	0.51 (1.34)	0.58 (1.55)	No	[144]
Moal et al.	Contact potentials	SKEMPI (1949)	0.73 (NR)	0.80 (1.25)	No	[146]

Relative performance of computational approaches to predict the effect of mutations on the binding free energy of protein-protein complexes
ML machine learning, *MD* molecular dynamics, *NR* not reported

An integrated computational approach was developed using structural information to understand the relation between the severity of phenotype, and the predicted effects of mutations on the stability of the pVHL protein and the change in affinity between pVHL and its protein partners (elongin B, elongin C and HIF-α peptide). The method, called Symphony (http://structure.bioc.cam.ac.uk/symphony), was able to predict the effects of mutations associated with RCC with high levels of sensitivity and specificity. A database of predictions for mutations not yet observed has also been developed.

mCSM-PPI has also been capable of giving a rationale for the affects of mutations on PPIs that are related to inherited RCC in other genes, including the P15-CDK6 complex. Figure 12.2a, b show examples of the affects of mutations on the stability of pVHL-HIF-α and P15-CDK6 complexes that were correctly identified by mCSM-PPI to dramatically reduce protein-protein affinity and potentially disrupt the complex.

Another successful application of computational predictors for understanding the mechanism of action of mutations in Mendelian diseases is the study of alkaptonuria (AKU). AKU is a rare, inherited metabolic disease caused by defective homogentisate 1,2-dioxygenase (HGD) as a result of mutations that disrupt its activity, many of them occurring in the protein-protein interfaces of its homo-hexamer (as the example shown in Fig. 12.2c). By using the predictions obtained by DUET [148] and mCSM-PPI, the mutations described in AKU were classified as belonging to one of three possible mechanism classes: protomer-destabilizing, PPI-destabilizing and active site mutations. Figure 12.2d depicts the distributions of these mutations on the structure of the human HGD.

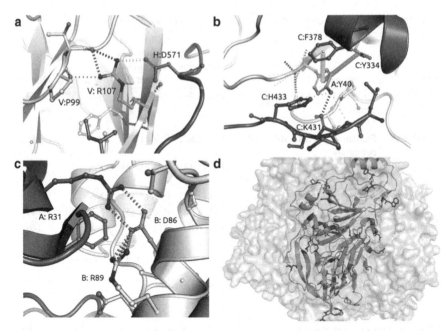

Fig. 12.2 Structural analysis of effects of mutations on PPIs and their role in cancer and Mendelian diseases. Panels (**a–c**) depict the interaction network made by important interface residues mutated in von Hippel-Lindau disease, alkaptonuria and renal cell carcinoma, respectively, whose effects were predicted to be highly destabilizing by mCSM-PPI. The chains of the binding partners are shown in dark grey. Panel (**a**) shows the interactions made by ARG107 in the interface between VHL and the HIF-2α peptide. Mutation to Proline will perturb local secondary structure and disrupt strong intra and inter-molecular hydrogen bonds and charged interactions. Panel (**b**) shows the TYR40 residue of the human HGD, which forms strong inter- and intra-molecular pi-pi interactions that are lost upon mutation to serine in AKU. Panel (**c**) shows the P15-CDK6 complex. Mutation of residue ASP86 on P15 to asparagine results in a loss of a charged interaction at the core of the interface. Panel (**d**) highlights a selection of AKU mutations within human HGD that mCSM-PPI predicts will reduce protein-protein affinity, leading to a loss of enzyme activity. These are spread across the extensive binding interface

12.7 Mutations & Interfaces: Hotspot Identification

Experimental and computational alanine scanning have been popular approaches to identifying amino acids that are critical for the formation of the complex, termed hotspot residues [112, 132, 133]. Robetta alanine scanning defines hotspot residues as those that upon mutation to alanine are predicted to decrease the binding energy by at least $\Delta\Delta G$ 1.0 kcal/mol.

Using mCSM-PPI, an alanine scan of 743 mutations within 19 different protein-protein complexes was performed to identify potential hotspot residues. The distribution of changes (Fig. 12.3) in binding free energy are consistent with the hypothesis that the loss of hotspot residues would have a significant effect

Fig. 12.3 Density
distribution of protein-protein
affinity change predictions by
mCSM-PPI for mutations that
were experimentally assigned
as hotspots. While the
majority of the neutral
mutations (*dotted-line curve*)
were predicted to have little
impact, most of the hotspots
(*solid-line curve*) were
predicted to have a significant
destabilizing effect on the
protein-protein complex

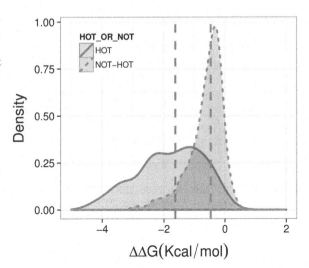

upon the affinity of two binding partners. The predicted change in binding energy
showed that mCSM-PPI predictions correlated strongly with the experimental data
($r > 0.7$). This indicates that mCSM-PPI could also be a powerful tool for hotspot
identification.

12.8   Examples of Success Using FBDD to Target PPI's

PPI's have been successfully modulated by compounds that mimic protein interaction elements, including proteomimetics [149], foldamers [150], peptide aptamers [151], antibodies [152, 153] and affibodies [154], where unfavorable pharmacokinetic properties are modified by the use of drug carriers or chemical modifications like PEGylation [155]. However, the development of more traditional pharmaceutical small molecule modulators, which remains highly desired, is proving a viable strategy, as demonstrated by several small molecule PPI inhibitors currently used therapeutically including the anti-HIV drug Maraviroc, an inhibitor of the CCR5-gp120 interaction, and Titrobifan, a glycoprotein IIb/IIIa inhibitor used in cardiovascular disease.

Two main resources store small molecule data modulating protein interfaces: the 2P2I database [156] dedicated to structural complexes of orthosteric PPI inhibitors of PPI and the TIMBAL database [157] that holds small molecule data for PPI modulators (inhibitors and stabilizers). Comparison of the original contents of the TIMBAL database [158] with known drugs and standard screening compounds revealed that small molecules disturbing protein assemblies were bigger, more lipophilic and with less polar features than the drugs and standard synthetic molecules. Analysis of the contacts these inhibitors made when the structure was

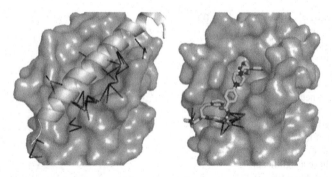

Fig. 12.4 Bcl-XL. *Left*: pdb code 2BZW, Bcl-XL (*surface representation*) bound to BAD (*cartoon representation*). *Right*: pdb code 2YXJ, Bcl-XL (*surface representation*) with the Abbott compound ABT-737 (*sticks representation*). Only polar contacts are shown

available corroborated that on average these small molecules were engaging mainly in hydrophobic contacts with the protein target.

These observations raised the question as to whether the lipophilicity is a requirement for binding to protein-protein interfaces or a reflection of common sins in drug discovery [159]. Comparison of protein complexes with PPI inhibitors, including synthetic and natural small molecules, small peptides and other proteins, highlighted the fact that protein complexes and natural molecules tend to interact with higher ratios of polar to non-polar contacts than synthetic small molecules. Contrasting the few cases where structures exist for both the protein-protein and the protein-PPI inhibitor complexes, synthetic small molecules were shown to miss available polar contact opportunities at the protein interface [160]. Figure 12.4 shows a graphical view of this concept.

When a structure is available, however, it can provide very useful insight and plays a crucial role in the development of fragment hits. The cytokine interleukin-2 (IL-2) induces T cell proliferation through binding to its heterotrimeric receptor. The structure of a small molecule inhibitor of this interaction identified by Hoffman-La Roche revealed that binding to IL-2 induced a significant conformational change to create a hydrophobic binding pocket that could accommodate the inhibitor [161]. This region overlapped with hotspot residues identified by alanine scanning mutagenesis [162, 163]. Based on this information, Wells and colleagues created a series of 11 cysteine mutants to identify small molecule inhibitors through tethering [161]. This identified a number of fragments that were shown to bind with sub-micromolar affinity. Medicinal chemistry was able to improve this affinity further to the low micromolar range. The crystal structures of the complexes, however, revealed two fragments bound to close but distinct sites. Linking these fragments together, they were able to achieve nanomolar inhibitors of the IL-2 interaction.

An example of the power of biophysical FBDD approaches to target a PPI is the development of inhibitors of the interaction between the human recombinase RAD51 and BRCA2 [52, 164]. Initial screening of a fragment library by thermal shift, followed by validation using NMR and X-ray crystallography resulted in the structures of approximately 80 fragments bound to RAD51, which disrupted the interaction with BRAC2. With structural information in hand, fragment growing or fragment linking can be employed to identify larger compounds from one or more fragment starting points. In general, as a fragment is expanded to make additional interactions in a fragment-based drug discovery campaign, affinities tend to be increased by 3–5 orders of magnitude [165–169]. The growth of the fragments bound to RAD51 was guided by the co-crystallized structures together with the structure of RAD51 in complex with the BRC4 region of BRAC2, and was able to improve the K_D from the mM to sub mM range [52]. More recently nanomolar affinities have been achieved by the Cambridge Group.

How can we tackle these challenging interfaces using chemistry that brings more polar specific contacts into play? On the one hand, interfaces where a flexible peptide binds to a continuous epitope in a concerted folding seem to offer more opportunities for "ligandability" [170] than preformed globular protein partners assemblies [103]. On the other hand, fragment-based approaches [171, 172] give pivotal advantages for these targets as they identify the hotspots by binding and yield less hydrophobic hits [173].

12.9   Final Thoughts

PPI's play a crucial role within the cell and their perturbation can lead to a range of diseases. They also present attractive and selective sites for drug development. Significant improvements in methodology have allowed the development of some highly selective modulators. Although targeting protein-protein interfaces still presents considerable technical challenges, as our understanding of these sites continues to expand, so too will our ability to modulate them selectively.

References

1. Lipinski CA, Lombardo F, Dominy BW et al (2001) Experimental and computational approaches to estimate solubility and permeability in drug discovery and development settings. Adv Drug Deliv Rev 46:3–26
2. Blundell T, Sibanda BL, Pearl L (1983) Three-dimensional structure, specificity and catalytic mechanism of renin. Nature 304:273–275
3. Foundling SI, Cooper J, Watson FE et al (1987) High resolution X-ray analyses of renin inhibitor-aspartic proteinase complexes. Nature 327:349–352
4. Lapatto R, Blundell T, Hemmings A et al (1989) X-ray analysis of HIV-1 proteinase at 2.7 A resolution confirms structural homology among retroviral enzymes. Nature 342:299–302
5. Dhanaraj V, Dealwis CG, Frazao C et al (1992) X-ray analyses of peptide-inhibitor complexes define the structural basis of specificity for human and mouse renins. Nature 357:466–472
6. Albiston AL, Morton CJ, Ng HL et al (2008) Identification and characterization of a new cognitive enhancer based on inhibition of insulin-regulated aminopeptidase. FASEB J 22:4209–4217

7. Ascher DB, Polekhina G, Parker MW (2012) Crystallization and preliminary X-ray diffraction analysis of human endoplasmic reticulum aminopeptidase 2. Acta Crystallogr Sect F: Struct Biol Cryst Commun 68:468–471
8. Chai SY, Yeatman HR, Parker MW et al (2008) Development of cognitive enhancers based on inhibition of insulin-regulated aminopeptidase. BMC Neurosci 9(Suppl 2):S14
9. Ye S, Chai SY, Lew RA et al (2008) Identification of modulating residues defining the catalytic cleft of insulin-regulated aminopeptidase. Biochem Cell Biol 86:251–261
10. Parker LJ, Ascher DB, Gao C et al (2012) Structural approaches to probing metal interaction with proteins. J Inorg Biochem 115:138–147
11. Parker LJ, Italiano LC, Morton CJ et al (2011) Studies of glutathione transferase P1-1 bound to a platinum(IV)-based anticancer compound reveal the molecular basis of its activation. Chemistry 17:7806–7816
12. Zhang J, Yang PL, Gray NS (2009) Targeting cancer with small molecule kinase inhibitors. Nat Rev Cancer 9:28–39
13. Congreve M, Carr R, Murray C et al (2003) A 'rule of three' for fragment-based lead discovery? Drug Discov Today 8:876–877
14. Erlanson DA, Braisted AC, Raphael DR et al (2000) Site-directed ligand discovery. Proc Natl Acad Sci U S A 97:9367–9372
15. Blundell TL, Jhoti H, Abell C (2002) High-throughput crystallography for lead discovery in drug design. Nat Rev Drug Discov 1:45–54
16. Congreve M, Murray CW, Blundell TL (2005) Structural biology and drug discovery. Drug Discov Today 10:895–907
17. Lepre CA, Moore JM, Peng JW (2004) Theory and applications of NMR-based screening in pharmaceutical research. Chem Rev 104:3641–3676
18. Murray CW, Blundell TL (2010) Structural biology in fragment-based drug design. Curr Opin Struct Biol 20:497–507
19. Cala O, Guilliere F, Krimm I (2014) NMR-based analysis of protein-ligand interactions. Anal Bioanal Chem 406:943–956
20. Hartshorn MJ, Murray CW, Cleasby A et al (2005) Fragment-based lead discovery using X-ray crystallography. J Med Chem 48:403–413
21. Caliandro R, Belviso DB, Aresta BM et al (2013) Protein crystallography and fragment-based drug design. Future Med Chem 5:1121–1140
22. Navratilova I, Hopkins AL (2010) Fragment screening by surface plasmon resonance. ACS Med Chem Lett 1:44–48
23. Navratilova I, Hopkins AL (2011) Emerging role of surface plasmon resonance in fragment-based drug discovery. Future Med Chem 3:1809–1820
24. Shepherd CA, Hopkins AL, Navratilova I (2014) Fragment screening by SPR and advanced application to GPCRs. Biol Prog Biophys Mol 116:113–123
25. Pantoliano MW, Petrella EC, Kwasnoski JD et al (2001) High-density miniaturized thermal shift assays as a general strategy for drug discovery. J Biomol Screen 6:429–440
26. Lo MC, Aulabaugh A, Jin G et al (2004) Evaluation of fluorescence-based thermal shift assays for hit identification in drug discovery. Anal Biochem 332:153–159
27. Kranz JK, Schalk-Hihi C (2011) Protein thermal shifts to identify low molecular weight fragments. Methods Enzymol 493:277–298
28. Leavitt S, Freire E (2001) Direct measurement of protein binding energetics by isothermal titration calorimetry. Curr Opin Struct Biol 11:560–566
29. Gozalbes R, Carbajo RJ, Pineda-Lucena A (2010) Contributions of computational chemistry and biophysical techniques to fragment-based drug discovery. Curr Med Chem 17: 1769–1794
30. Ladbury JE, Klebe G, Freire E (2010) Adding calorimetric data to decision making in lead discovery: a hot tip. Nat Rev Drug Discov 9:23–27
31. Valkov E, Sharpe T, Marsh M et al (2012) Targeting protein-protein interactions and fragment-based drug discovery. Top Curr Chem 317:145–179

32. Whittle PJ, Blundell TL (1994) Protein structure–based drug design. Annu Rev Biophys Biomol Struct 23:349–375
33. Surade S, Blundell TL (2012) Structural biology and drug discovery of difficult targets: the limits of ligandability. Chem Biol 19:42–50
34. Blundell TL, Srinivasan N (1996) Symmetry, stability, and dynamics of multidomain and multicomponent protein systems. Proc Natl Acad Sci U S A 93:14243–14248
35. Pellegrini L, Burke DF, von Delft F et al (2000) Crystal structure of fibroblast growth factor receptor ectodomain bound to ligand and heparin. Nature 407:1029–1034
36. Harmer NJ, Ilag LL, Mulloy B et al (2004) Towards a resolution of the stoichiometry of the fibroblast growth factor (FGF)-FGF receptor-heparin complex. J Mol Biol 339:821–834
37. Robinson CJ, Harmer NJ, Goodger SJ et al (2005) Cooperative dimerization of fibroblast growth factor 1 (FGF1) upon a single heparin saccharide may drive the formation of 2:2:1 FGF1.FGFR2c.heparin ternary complexes. J Biol Chem 280:42274–42282
38. Brown A, Robinson CJ, Gallagher JT et al (2013) Cooperative heparin-mediated oligomerization of fibroblast growth factor-1 (FGF1) precedes recruitment of FGFR2 to ternary complexes. Biophys J 104:1720–1730
39. Chirgadze DY, Hepple JP, Zhou H et al (1999) Crystal structure of the NK1 fragment of HGF/SF suggests a novel mode for growth factor dimerization and receptor binding. Nat Struct Biol 6:72–79
40. Gherardi E, Youles ME, Miguel RN et al (2003) Functional map and domain structure of MET, the product of the c-met protooncogene and receptor for hepatocyte growth factor/scatter factor. Proc Natl Acad Sci U S A 100:12039–12044
41. Gherardi E, Sandin S, Petoukhov MV et al (2006) Structural basis of hepatocyte growth factor/scatter factor and MET signalling. Proc Natl Acad Sci U S A 103:4046–4051
42. Higueruelo AP, Jubb H, Blundell TL (2013) Protein-protein interactions as druggable targets: recent technological advances. Curr Opin Pharmacol 13:791–796
43. Herbert C, Schieborr U, Saxena K et al (2013) Molecular mechanism of SSR128129E, an extracellularly acting, small-molecule, allosteric inhibitor of FGF receptor signaling. Cancer Cell 23:489–501
44. Bolanos-Garcia VM, Wu Q, Ochi T et al (2012) Spatial and temporal organization of multi-protein assemblies: achieving sensitive control in information-rich cell-regulatory systems. Philos Transact A Math Phys Eng Sci 370:3023–3039
45. Sibanda BL, Critchlow SE, Begun J et al (2001) Crystal structure of an Xrcc4-DNA ligase IV complex. Nat Struct Biol 8:1015–1019
46. Sibanda BL, Chirgadze DY, Blundell TL (2010) Crystal structure of DNA-PKcs reveals a large open-ring cradle comprised of HEAT repeats. Nature 463:118–121
47. Singleton BK, Torres-Arzayus MI, Rottinghaus ST et al (1999) The C terminus of Ku80 activates the DNA-dependent protein kinase catalytic subunit. Mol Cell Biol 19:3267–3277
48. Gell D, Jackson SP (1999) Mapping of protein-protein interactions within the DNA-dependent protein kinase complex. Nucleic Acids Res 27:3494–3502
49. Nick McElhinny SA, Snowden CM, McCarville J et al (2000) Ku recruits the XRCC4-ligase IV complex to DNA ends. Mol Cell Biol 20:2996–3003
50. Yano K, Morotomi-Yano K, Wang SY et al (2008) Ku recruits XLF to DNA double-strand breaks. EMBO Rep 9:91–96
51. Sasaki K, Dockerill S, Adamiak DA et al (1975) X-ray analysis of glucagon and its relationship to receptor binding. Nature 257:751–757
52. Pellegrini L, Yu DS, Lo T et al (2002) Insights into DNA recombination from the structure of a RAD51-BRCA2 complex. Nature 420:287–293
53. Dyson HJ, Wright PE (2002) Coupling of folding and binding for unstructured proteins. Curr Opin Struct Biol 12:54–60
54. Hernandez H, Robinson CV (2007) Determining the stoichiometry and interactions of macromolecular assemblies from mass spectrometry. Nat Protoc 2:715–726
55. Sharon M, Robinson CV (2007) The role of mass spectrometry in structure elucidation of dynamic protein complexes. Annu Rev Biochem 76:167–193

56. Ascher DB, Cromer BA, Morton CJ et al (2011) Regulation of insulin-regulated membrane aminopeptidase activity by its C-terminal domain. Biochemistry 50:2611–2622
57. Ascher DB, Wielens J, Nero TL et al (2014) Potent hepatitis C inhibitors bind directly to NS5A and reduce its affinity for RNA. Sci Rep 4:4765
58. Polekhina G, Ascher DB, Kok SF et al (2013) Structure of the N-terminal domain of human thioredoxin-interacting protein. Acta Crystallogr D Biol Crystallogr 69:333–344
59. Rambo RP, Tainer JA (2013) Accurate assessment of mass, models and resolution by small-angle scattering. Nature 496:477–481
60. Koch MH, Vachette P, Svergun DI (2003) Small-angle scattering: a view on the properties, structures and structural changes of biological macromolecules in solution. Q Rev Biophys 36:147–227
61. Putnam CD, Hammel M, Hura GL et al (2007) X-ray solution scattering (SAXS) combined with crystallography and computation: defining accurate macromolecular structures, conformations and assemblies in solution. Q Rev Biophys 40:191–285
62. Hammel M, Yu Y, Fang S et al (2010) XLF regulates filament architecture of the XRCC4.ligase IV complex. Structure 18:1431–1442
63. Wong W, Bai XC, Brown A et al (2014) Cryo-EM structure of the Plasmodium falciparum 80S ribosome bound to the anti-protozoan drug emetine. Elife 3:e03080
64. Davis AJ, Chen BP, Chen DJ (2014) DNA-PK: a dynamic enzyme in a versatile DSB repair pathway. DNA Repair (Amst) 17:21–29
65. Yang H, Rudge DG, Koos JD et al (2013) mTOR kinase structure, mechanism and regulation. Nature 497:217–223
66. Boskovic J, Rivera-Calzada A, Maman JD et al (2003) Visualization of DNA-induced conformational changes in the DNA repair kinase DNA-PKcs. EMBO J 22:5875–5882
67. Chiu CY, Cary RB, Chen DJ et al (1998) Cryo-EM imaging of the catalytic subunit of the DNA-dependent protein kinase. J Mol Biol 284:1075–1081
68. Leuther KK, Hammarsten O, Kornberg RD et al (1999) Structure of DNA-dependent protein kinase: implications for its regulation by DNA. EMBO J 18:1114–1123
69. Williams DR, Lee KJ, Shi J et al (2008) Cryo-EM structure of the DNA-dependent protein kinase catalytic subunit at subnanometer resolution reveals alpha helices and insight into DNA binding. Structure 16:468–477
70. Hammel M, Yu Y, Mahaney BL et al (2010) Ku and DNA-dependent protein kinase dynamic conformations and assembly regulate DNA binding and the initial non-homologous end joining complex. J Biol Chem 285:1414–1423
71. Morris EP, Rivera-Calzada A, da Fonseca PC et al (2011) Evidence for a remodelling of DNA-PK upon autophosphorylation from electron microscopy studies. Nucleic Acids Res 39:5757–5767
72. Spagnolo L, Rivera-Calzada A, Pearl LH et al (2006) Three-dimensional structure of the human DNA-PKcs/Ku70/Ku80 complex assembled on DNA and its implications for DNA DSB repair. Mol Cell 22:511–519
73. Cary RB, Peterson SR, Wang J et al (1997) DNA looping by Ku and the DNA-dependent protein kinase. Proc Natl Acad Sci U S A 94:4267–4272
74. Yaneva M, Kowalewski T, Lieber MR (1997) Interaction of DNA-dependent protein kinase with DNA and with Ku: biochemical and atomic-force microscopy studies. EMBO J 16: 5098–5112
75. Ochi T, Wu Q, Blundell TL (2014) The spatial organization of non-homologous end joining: from bridging to end joining. DNA Repair (Amst) 17:98–109
76. Andres SN, Vergnes A, Ristic D et al (2012) A human XRCC4-XLF complex bridges DNA. Nucleic Acids Res 40:1868–1878
77. Hammel M, Rey M, Yu Y et al (2011) XRCC4 protein interactions with XRCC4-like factor (XLF) create an extended grooved scaffold for DNA ligation and double strand break repair. J Biol Chem 286:32638–32650
78. Ropars V, Drevet P, Legrand P et al (2011) Structural characterization of filaments formed by human Xrcc4-Cernunnos/XLF complex involved in nonhomologous DNA end-joining. Proc Natl Acad Sci U S A 108:12663–12668

79. Wu Q, Ochi T, Matak-Vinkovic D et al (2011) Non-homologous end-joining partners in a helical dance: structural studies of XLF-XRCC4 interactions. Biochem Soc Trans 39: 1387–1392, suppl 2 p following 1392
80. Ochi T, Wu Q, Chirgadze DY et al (2012) Structural insights into the role of domain flexibility in human DNA ligase IV. Structure 20:1212–1222
81. Williams GJ, Hammel M, Radhakrishnan SK et al (2014) Structural insights into NHEJ: building up an integrated picture of the dynamic DSB repair super complex, one component and interaction at a time. DNA Repair (Amst) 17:110–120
82. Critchlow SE, Bowater RP, Jackson SP (1997) Mammalian DNA double-strand break repair protein XRCC4 interacts with DNA ligase IV. Curr Biol 7:588–598
83. Grawunder U, Zimmer D, Lieber MR (1998) DNA ligase IV binds to XRCC4 via a motif located between rather than within its BRCT domains. Curr Biol 8:873–876
84. Recuero-Checa MA, Dore AS, Arias-Palomo E et al (2009) Electron microscopy of Xrcc4 and the DNA ligase IV-Xrcc4 DNA repair complex. DNA Repair (Amst) 8:1380–1389
85. Ochi T, Gu X, Blundell TL (2013) Structure of the catalytic region of DNA ligase IV in complex with an Artemis fragment sheds light on double-strand break repair. Structure 21:672–679
86. Berman HM, Westbrook J, Feng Z et al (2000) The protein data bank. Nucleic Acids Res 28:235–242
87. Dou Y, Baisnee PF, Pollastri G et al (2004) ICBS: a database of interactions between protein chains mediated by beta-sheet formation. Bioinformatics 20:2767–2777
88. Lo A, Cheng CW, Chiu YY et al (2011) TMPad: an integrated structural database for helix-packing folds in transmembrane proteins. Nucleic Acids Res 39:D347–D355
89. Mosca R, Ceol A, Stein A et al (2014) 3did: a catalog of domain-based interactions of known three-dimensional structure. Nucleic Acids Res 42:D374–D379
90. Huang Z, Zhu L, Cao Y et al (2011) ASD: a comprehensive database of allosteric proteins and modulators. Nucleic Acids Res 39:D663–D669
91. Huang Z, Mou L, Shen Q et al (2014) ASD v2.0: updated content and novel features focusing on allosteric regulation. Nucleic Acids Res 42:D510–D516
92. Bickerton GR, Higueruelo AP, Blundell TL (2011) Comprehensive, atomic-level characterization of structurally characterized protein-protein interactions: the PICCOLO database. BMC Bioinf 12:313
93. Lee S, Blundell TL (2009) BIPA: a database for protein-nucleic acid interaction in 3D structures. Bioinformatics 25:1559–1560
94. Schreyer A, Blundell T (2009) CREDO: a protein-ligand interaction database for drug discovery. Chem Biol Drug Des 73:157–167
95. Schreyer AM, Blundell TL (2013) CREDO: a structural interactomics database for drug discovery. Database (Oxford) 2013:bat049
96. Hamosh A, Scott AF, Amberger JS et al (2005) Online Mendelian Inheritance in Man (OMIM), a knowledgebase of human genes and genetic disorders. Nucleic Acids Res 33:D514–D517
97. Forbes SA, Bhamra G, Bamford S et al (2008) The catalogue of somatic mutations in cancer (COSMIC). Curr Protoc Hum Genet, Chapter 10:Unit 10 11
98. Apweiler R, Bairoch A, Wu CH et al (2004) UniProt: the Universal Protein knowledgebase. Nucleic Acids Res 32:D115–D119
99. Hubbard T, Barker D, Birney E et al (2002) The Ensembl genome database project. Nucleic Acids Res 30:38–41
100. Gaulton A, Bellis LJ, Bento AP et al (2011) ChEMBL: a large-scale bioactivity database for drug discovery. Nucleic Acids Res 40:D1100–D1107
101. Lewell XQ, Judd DB, Watson SP et al (1998) RECAP-Retrosynthetic combinatorial analysis procedure: a powerful new technique for identifying privileged molecular fragments with useful applications in combinatorial chemistry. J Chem Inf Comput Sci 38:511–522
102. Deng Z, Chuaqui C, Singh J (2004) Structural interaction fingerprint (SIFt): a novel method for analyzing three-dimensional protein-ligand binding interactions. J Med Chem 47: 337–344

103. Blundell TL, Sibanda BL, Montalvao RW et al (2006) Structural biology and bioinformatics in drug design: opportunities and challenges for target identification and lead discovery. Philos Trans R Soc Lond B Biol Sci 361:413–423

104. Nair SK, Burley SK (2003) X-ray structures of Myc-Max and Mad-Max recognizing DNA. Molecular bases of regulation by proto-oncogenic transcription factors. Cell 112:193–205

105. Fletcher S, Hamilton AD (2006) Targeting protein-protein interactions by rational design: mimicry of protein surfaces. J R Soc Interface 3:215–233

106. Jones S, Thornton JM (1996) Principles of protein-protein interactions. Proc Natl Acad Sci U S A 93:13–20

107. Arkin MR, Tang Y, Wells JA (2014) Small-molecule inhibitors of protein-protein interactions: progressing toward the reality. Chem Biol 21:1102–1114

108. Cooper A (1999) Thermodynamic analysis of biomolecular interactions. Curr Opin Chem Biol 3:557–563

109. Breiten B, Lockett MR, Sherman W et al (2013) Water networks contribute to enthalpy/entropy compensation in protein-ligand binding. J Am Chem Soc 135:15579–15584

110. Fuller JC, Burgoyne NJ, Jackson RM (2009) Predicting druggable binding sites at the protein-protein interface. Drug Discov Today 14:155–161

111. Li X, Keskin O, Ma B et al (2004) Protein-protein interactions: hot spots and structurally conserved residues often locate in complemented pockets that pre-organized in the unbound states: implications for docking. J Mol Biol 344:781–795

112. Clackson T, Wells JA (1995) A hot spot of binding energy in a hormone-receptor interface. Science 267:383–386

113. Bogan AA, Thorn KS (1998) Anatomy of hot spots in protein interfaces. J Mol Biol 280:1–9

114. Rajamani D, Thiel S, Vajda S et al (2004) Anchor residues in protein-protein interactions. Proc Natl Acad Sci U S A 101:11287–11292

115. Ben-Shimon A, Eisenstein M (2010) Computational mapping of anchoring spots on protein surfaces. J Mol Biol 402:259–277

116. Meireles LM, Domling AS, Camacho CJ (2010) ANCHOR: a web server and database for analysis of protein-protein interaction binding pockets for drug discovery. Nucleic Acids Res 38:W407–W411

117. Koes DR, Camacho CJ (2012) PocketQuery: protein-protein interaction inhibitor starting points from protein-protein interaction structure. Nucleic Acids Res 40:W387–W392

118. London N, Movshovitz-Attias D, Schueler-Furman O (2010) The structural basis of peptide-protein binding strategies. Structure 18:188–199

119. London N, Raveh B, Movshovitz-Attias D et al (2010) Can self-inhibitory peptides be derived from the interfaces of globular protein-protein interactions? Proteins 78:3140–3149

120. London N, Raveh B, Schueler-Furman O (2013) Druggable protein-protein interactions–from hot spots to hot segments. Curr Opin Chem Biol 17:952–959

121. Pommier Y, Marchand C (2012) Interfacial inhibitors: targeting macromolecular complexes. Nat Rev Drug Discov 11:25–36

122. Gao M, Skolnick J (2012) The distribution of ligand-binding pockets around protein-protein interfaces suggests a general mechanism for pocket formation. Proc Natl Acad Sci U S A 109:3784–3789

123. Topham CM, Srinivasan N, Blundell TL (1997) Prediction of the stability of protein mutants based on structural environment-dependent amino acid substitution and propensity tables. Protein Eng 10:7–21

124. Worth CL, Preissner R, Blundell TL (2011) SDM–a server for predicting effects of mutations on protein stability and malfunction. Nucleic Acids Res 39:W215–W222

125. Overington J, Donnelly D, Johnson MS et al (1992) Environment-specific amino acid substitution tables: tertiary templates and prediction of protein folds. Protein Sci 1:216–226

126. Kucukkal TG, Yang Y, Chapman SC et al (2014) Computational and experimental approaches to reveal the effects of single nucleotide polymorphisms with respect to disease diagnostics. Int J Mol Sci 15:9670–9717

127. Pires DE, Ascher DB, Blundell TL (2014) mCSM: predicting the effects of mutations in proteins using graph-based signatures. Bioinformatics 30:335–342

128. da Silveira CH, Pires DE, Minardi RC et al (2009) Protein cutoff scanning: a comparative analysis of cutoff dependent and cutoff free methods for prospecting contacts in proteins. Proteins 74:727–743

129. Pires DE, de Melo-Minardi RC, Ados Santos M et al (2011) Cutoff Scanning Matrix (CSM): structural classification and function prediction by protein inter-residue distance patterns. BMC Genomics 12(Suppl 4):S12

130. Pires DE, de Melo-Minardi RC, da Silveira CH et al (2013) aCSM: noise-free graph-based signatures to large-scale receptor-based ligand prediction. Bioinformatics 29:855–861

131. Guerois R, Nielsen JE, Serrano L (2002) Predicting changes in the stability of proteins and protein complexes: a study of more than 1000 mutations. J Mol Biol 320:369–387

132. Kortemme T, Baker D (2002) A simple physical model for binding energy hot spots in protein-protein complexes. Proc Natl Acad Sci U S A 99:14116–14121

133. Kortemme T, Kim DE, Baker D (2004) Computational alanine scanning of protein-protein interfaces. Sci STKE 2004:pl2

134. Huo S, Massova I, Kollman PA (2002) Computational alanine scanning of the 1:1 human growth hormone-receptor complex. J Comput Chem 23:15–27

135. Gouda H, Kuntz ID, Case DA et al (2003) Free energy calculations for theophylline binding to an RNA aptamer: comparison of MM-PBSA and thermodynamic integration methods. Biopolymers 68:16–34

136. Kollman PA, Massova I, Reyes C et al (2000) Calculating structures and free energies of complex molecules: combining molecular mechanics and continuum models. Acc Chem Res 33:889–897

137. Kollman P (1993) Free energy calculations: applications to chemical and biochemical phenomena. Chem Rev 93:2395–2417

138. Kong X, Brooks C (1996) Lambda-dynamics: a new approach to free energy calculations. J Chem Phys 105:2414–2423

139. Moreira I, Fernandes P, Ramos M (2007) Unravelling Hot Spots: a comprehensive computational mutagenesis study. Theor Chem Accounts 117:99–113

140. Moreira IS, Fernandes PA, Ramos MJ (2005) Accuracy of the numerical solution of the Poisson–Boltzmann equation. J Mol Struct THEOCHEM 729:11–18

141. Massova I, Kollman PA (1999) Computational alanine scanning to probe protein–protein interactions: a novel approach to evaluate binding free energies. J Am Chem Soc 121:8133–8143

142. Moal IH, Fernandez-Recio J (2012) SKEMPI: a structural kinetic and energetic database of mutant protein interactions and its use in empirical models. Bioinformatics 28:2600–2607

143. Dehouck Y, Kwasigroch JM, Rooman M et al (2013) BeAtMuSiC: prediction of changes in protein-protein binding affinity on mutations. Nucleic Acids Res 41:W333–W339

144. Dourado DF, Flores SC (2014) A multiscale approach to predicting affinity changes in protein-protein interfaces. Proteins 82:2681–2690

145. Li M, Petukh M, Alexov E et al (2014) Predicting the impact of missense mutations on protein-protein binding affinity. J Chem Theory Comput 10:1770–1780

146. Moal IH, Fernandez-Recio J (2013) Intermolecular contact potentials for protein–protein interactions extracted from binding free energy changes upon mutation. J Chem Theory Comput 9:3715–3727

147. Gossage L, Pires DE, Olivera-Nappa A et al (2014) An integrated computational approach can classify VHL missense mutations according to risk of clear cell renal carcinoma. Hum Mol Genet 23:5976–5988

148. Pires DE, Ascher DB, Blundell TL (2014) DUET: a server for predicting effects of mutations on protein stability using an integrated computational approach. Nucleic Acids Res 42:W314–W319

149. Fletcher S, Hamilton AD (2005) Protein surface recognition and proteomimetics: mimics of protein surface structure and function. Curr Opin Chem Biol 9:632–638
150. Wilson AJ (2009) Inhibition of protein-protein interactions using designed molecules. Chem Soc Rev 38:3289–3300
151. Buerger C, Groner B (2003) Bifunctional recombinant proteins in cancer therapy: cell penetrating peptide aptamers as inhibitors of growth factor signaling. J Cancer Res Clin Oncol 129:669–675
152. Seidah NG (2013) Proprotein convertase subtilisin kexin 9 (PCSK9) inhibitors in the treatment of hypercholesterolemia and other pathologies. Curr Pharm Des 19:3161–3172
153. Traczewski P, Rudnicka L (2011) Treatment of systemic lupus erythematosus with epratuzumab. Br J Clin Pharmacol 71:175–182
154. Nord K, Gunneriusson E, Ringdahl J et al (1997) Binding proteins selected from combinatorial libraries of an alpha-helical bacterial receptor domain. Nat Biotechnol 15:772–777
155. Kaminskas LM, Ascher DB, McLeod VM et al (2013) PEGylation of interferon alpha2 improves lymphatic exposure after subcutaneous and intravenous administration and improves antitumour efficacy against lymphatic breast cancer metastases. J Control Release 168:200–208
156. Basse MJ, Betzi S, Bourgeas R et al (2013) 2P2Idb: a structural database dedicated to orthosteric modulation of protein-protein interactions. Nucleic Acids Res 41:D824–D827
157. Higuereulo AP, Jubb H, Blundell TL (2013) TIMBAL v2: update of a database holding small molecules modulating protein-protein interactions. Database (Oxford) 2013:bat039
158. Higuereulo AP, Schreyer A, Bickerton GRJ et al (2009) Atomic interactions and profile of small molecules disrupting protein-protein interfaces: the TIMBAL database. Chem Biol Drug Des 74:457–467
159. Hann MM (2011) Molecular obesity, potency and other addictions in drug discovery. MedChemComm 2:349–355
160. Higuereulo AP, Schreyer A, Bickerton GRJ et al (2012) What can we learn from the evolution of protein-ligand interactions to aid the design of new therapeutics? PLoS One 7:e51742
161. Arkin MR, Randal M, DeLano WL et al (2003) Binding of small molecules to an adaptive protein-protein interface. Proc Natl Acad Sci U S A 100:1603–1608
162. Sauve K, Nachman M, Spence C et al (1991) Localization in human interleukin 2 of the binding site to the alpha chain (p55) of the interleukin 2 receptor. Proc Natl Acad Sci U S A 88:4636–4640
163. Zurawski SM, Vega F Jr, Doyle EL et al (1993) Definition and spatial location of mouse interleukin-2 residues that interact with its heterotrimeric receptor. EMBO J 12:5113–5119
164. Scott DE, Ehebauer MT, Pukala T et al (2013) Using a fragment-based approach to target protein-protein interactions. Chembiochem 14:332–342
165. Wyatt PG, Woodhead AJ, Berdini V et al (2008) Identification of N-(4-piperidinyl)-4-(2,6-dichlorobenzoylamino)-1H-pyrazole-3-carboxamide (AT7519), a novel cyclin dependent kinase inhibitor using fragment-based X-ray crystallography and structure based drug design. J Med Chem 51:4986–4999
166. Howard S, Berdini V, Boulstridge JA et al (2009) Fragment-based discovery of the pyrazol-4-yl urea (AT9283), a multitargeted kinase inhibitor with potent aurora kinase activity. J Med Chem 52:379–388
167. Frederickson M, Callaghan O, Chessari G et al (2008) Fragment-based discovery of mexiletine derivatives as orally bioavailable inhibitors of urokinase-type plasminogen activator. J Med Chem 51:183–186
168. Antonysamy SS, Aubol B, Blaney J et al (2008) Fragment-based discovery of hepatitis C virus NS5b RNA polymerase inhibitors. Bioorg Med Chem Lett 18:2990–29995
169. Antonysamy S, Hirst G, Park F et al (2009) Fragment-based discovery of JAK-2 inhibitors. Bioorg Med Chem Lett 19:279–282

170. Edfeldt FNB, Folmer RHA, Breeze AL (2011) Fragment screening to predict druggability (ligandability) and lead discovery success. Drug Discov Today 16:284–287
171. Winter A, Higueruelo P et al (2012) Biophysical and computational fragment-based approaches to targeting protein-protein interactions: applications in structure-guided drug discovery. Q Rev Biophys 45:383–426
172. Jubb H, Higueruelo A, Winter A et al (2012) Structural biology and drug discovery for protein–protein interactions. Trends Pharmacol Sci 33:241–248
173. Keserü GM, Makara GM (2009) The influence of lead discovery strategies on the properties of drug candidates. Nat Rev Drug Discov 8:203–212

Chapter 13
Achieving High Quality Ligand Chemistry in Protein-Ligand Crystal Structures for Drug Design

Oliver S. Smart and Gérard Bricogne

Abstract The production of an X-ray crystal structure for a protein-ligand complex involves many steps, encompassing experimental and computational crystallography as well as chemoinformatics and computational chemistry. Using examples taken from the PDB, we show how a mistake made in any of these steps adversely affects the quality of the resulting structure, including that of the ligand. Procedures to assess the reliability of a ligand in a protein-ligand crystal structure are described. The merits of different responses to the identification of a problematic ligand structure in the PDB are examined. It is proposed that the best course of action is to cooperate with authors of the PDB entry and to deposit a corrected structure to replace the original. Two detailed examples of this process are provided by the deposition of improvements to PDB entries 1BYK and 1PMQ with their original depositors.

13.1 Introduction

This chapter looks at the steps necessary to produce a crystal structure of a protein-ligand complex with high-quality ligand placement. We will also look at ways of assessing the reliability of the ligand in such structures, whether taken from the Protein Data Bank (PDB) [1] or supplied by a colleague. The chapter is accompanied by two workshop practical sessions given at the Erice International School of Crystallography [2].

O.S. Smart (✉)
Global Phasing Ltd., Sheraton House, Castle Park, Cambridge CB3 0AX, UK

SmartSci Limited, St John's Innovation Centre, Cowley Road, Cambridge CB4 0WS, UK
e-mail: osmart@globalphasing.com

G. Bricogne
Global Phasing Ltd., Sheraton House, Castle Park, Cambridge CB3 0AX, UK

© Springer Science+Business Media Dordrecht 2015
G. Scapin et al. (eds.), *Multifaceted Roles of Crystallography in Modern Drug Discovery*, NATO Science for Peace and Security Series A: Chemistry and Biology, DOI 10.1007/978-94-017-9719-1_13

165

Crystal structures of protein-ligand complexes play a crucial role in structure-guided drug design [3]: they are used to understand protein structure-function relationships as well as for training and/or validating ligand docking programs [3–5]. Achieving reliable ligand placement in these structures is therefore of the utmost importance.

Producing a crystal structure for a ligand-soak experiment on a protein for which a complete X-ray structure of the ligand-free "apo" protein is already available typically involves:

(a) Experimental X-ray data collection from a protein crystal with the ligand soaked or co-crystallized, using a synchrotron beamline or an in-house diffractometer.

(b) Data processing and integration to give space group, unit cell and structure factor amplitudes (SF).

(c) Molecular replacement to optimally reposition the protein model for the cell and SF from (b) by rigid-body movements.

(d) Initial refinement of model from (c) without a ligand.

(e) Assessment of whether the difference electron density (ED) for the model from (d) warrants attempting to place a ligand.

(f) Produce a molecular model and a restraint dictionary for the ligand.

(g) Fit the model of ligand (f) into difference density and protein model from (d).

(h) Refinement of combined protein and ligand model.

(i) Assessment of refined protein-ligand complex (h).

(j) If assessment shows issues then rebuild/refit protein, ligand and/or solvent and back to step (h).

(k) Deposition of the structure model, SF, maps and validation data to an in-house database (or the PDB).

Most of these steps can be automated into a structure determination pipeline – for instance steps (b) to (i) are tackled by the Global Phasing tool PIPEDREAM [6]. A mistake made in any of these steps will adversely affect the quality of the resulting structure, including that of the ligand. To exemplify this we will examine a number of structures taken from the PDB. The PDB [1] is a databank of "complete" structures and provides a great resource for looking at mistakes made in solving protein-ligand complexes and for improving procedures so as to avoid such issues in the future [7]. This is particularly important both for the developers and for the users of automated pipelines.

13.1.1 Validation of the Ligand in the Crystal Structure of a Protein-Ligand Complex

It is important for the user of a protein-ligand structure to be able to assess the reliability of its ligand(s). For this purpose we have developed the BUSTER-REPORT program [8]. To use this tool BUSTER [9] is first run to produce ED maps or to refine the structure in question, and then BUSTER-REPORT will analyze results providing an HTML page that reports on:

- The X-ray data using the BUSTER reciprocal space correlation coefficients (RecSCC) plot. The RecSCC plot allows the detection of problems such as ice-ring contamination, anisotropic diffraction and incomplete data collection. For details see:
  http://www.globalphasing.com/buster/wiki/index.cgi?BusterReport.
- The usual statistics R_{work} and R_{free} as indicators of the overall progress and final performance of the refinement process in fitting the experimental X-ray data.
- MolProbity evaluation of protein geometry, including Ramachandran plots [10].

In addition BUSTER-REPORT provides reports for each ligand in the model, giving:

- Pictures of the ED around the ligand. These are provided as animated GIFs to aid visualization. The presence of large amounts of difference density around a ligand is a matter of concern (Fig. 13.1a).
- The real space correlation coefficient (CC) of the ligand, which provides an overall measure of the agreement between the $2F_o$-F_c ED and the molecular model of the ligand. CC values below 0.8 are a prompt to reconsider the ligand placement.
- The average and maximum B-factor for ligand atoms. The B-factors are adjusted in refinement and describe the degree to which the ED is spread out. High ligand B-factors are often an indication of problematic placement, unless a degree of local disorder is made plausible by the ligand's environment, e.g. its proximity to the solvent boundary.
- The results of MOGUL on the geometry of the ligand. The MOGUL [11] program is a tool that facilitates searching the Cambridge Structural Database of small-molecule organic and metal-organic crystal structures (CSD) [12] for geometric information relevant to a given ligand. MOGUL will rapidly analyze bond lengths, bond angles and most dihedral angles by finding CSD entries that contain similar chemical groups. In addition it provides data for many five and six-membered rings checking whether the ring pucker is similar to that found in related CSD entries. BUSTER-REPORT presents the results of this evaluation of geometric quality by means of colored 2D diagrams of each ligand (Fig. 13.1b). Dihedral angles and ring scores are the most useful as metrics for validation, particularly if a GRADE [13] restraint dictionary is used in the refinement.

Although BUSTER-REPORT provides much useful information, it is best used together with direct visualization of the model and ED maps using COOT. This also gives an assessment of whether the ligand placement makes sense in terms of protein-ligand interactions. In general, correctly placed ligands will tend to form hydrogen-bond contacts to neighboring protein or solvent atoms as well as placing hydrophobic groups into hydrophobic environments.

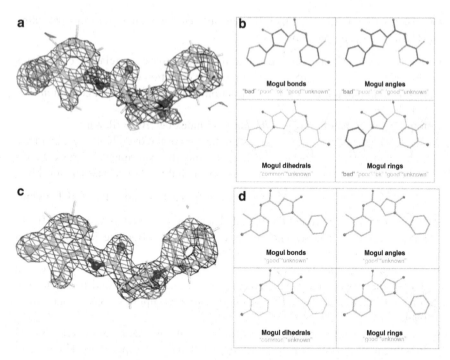

Fig. 13.1 PDB entry 2H7P [29] (1.86 Å resolution). Panel (**a**) BUSTER [9] maps show considerable difference density for the pyrrolidine carboxamide ligand as modeled in the PDB (**b**) MOGUL [11] validation measures show that the ligand has issues with bond lengths and angles and with ring puckers. After re-refinement with BUSTER using a refinement dictionary produced by GRADE [13] the fit to electron density is greatly improved (**c**) and no problems are found by MOGUL (**d**). All analysis and images are produced by BUSTER-REPORT [8]. The $2F_o$-F_c ED map is shown in *grey* at a contour-level of 1.3 rmsd. The F_o-F_c difference map is contoured at ± 3.0 rmsd and shown in *green* for positive difference density and *red* for negative difference density. The full BUSTER-REPORT output is available from the introductory workshop practical available on-line [2]. After seeing this analysis, Stroud and co-workers have deposited a corrected structure 4TZT into the PDB that has good ligand geometry and good fit to ED (Color figure online)

13.1.2 *Electron Density*

Examination of the electron density (ED) maps forms a crucial part of assessing whether a ligand in a protein-ligand complex can be relied upon. ED maps are produced by the program are used to refine the structure. During the refinement process the maps will periodically be examined by the crystallographer using the COOT program [14]. Agreement between the experimental model of the protein, ligands and solvent molecules is assessed, and the model is adjusted as necessary, for instance by moving a protein side-chain or by placing water molecules into yet unmodelled density. Automated tools are increasingly used to help with the building process, but human examination and intervention are still normally necessary.

The ED maps at the end of refinement and model building can be seen to be as important as the refined model itself in reporting the result. It would be particularly useful to have access to the actual maps that the authors examined and interpreted in their work, in the concise form of their Fourier coefficients (i.e. amplitudes and phases). Unfortunately, these coefficients are not currently captured in a routine manner by the PDB deposition process and are seldom available from the archive itself. The Electron Density Server (EDS) at Uppsala [15] provides maps recalculated with the REFMAC [16] refinement program for PDB entries where this is possible. EDS is a valuable resource for users of protein-ligand complexes from the PDB that enables rapid retrieval of the ED maps for most PDB depositions. Alternatively, BUSTER [9] includes tools that, for any given PDB code, will rapidly download data, calculate maps and provide a BUSTER-REPORT [8] analysis of the structure. The BUSTER maps can then be inspected using BUSTER-REPORT or displayed using COOT [14].

The $2F_o$-F_c map indicates where ED is to be found according to the experimental X-ray data and the current refined atomic model. The F_o-F_c difference map indicates regions where the current model fails to place sufficient electrons (positive difference, normally shown in green) or places too many electrons (negative difference, normally shown in red). As shown in Fig. 13.1c, re-refinement and/or rebuilding an incorrect model will tend to move atoms into the middle of $2F_o$-F_c density and will reduce the amount of difference density. It should be noted that ED maps are not fixed: they generally improve as refinement and model building proceed. This is because as the model becomes more exact, the phases derived from it become more accurate, which in turn results in more accurate maps where more features become interpretable. The difference maps then become more sensitive and better able to highlight further unmodelled density or necessary corrections to the model.

13.1.3 The Importance of the X-ray Data Resolution Limit

The crystal structure of a protein-ligand complex is the result of an experiment where data are collected from a crystal of the protein soaked in, or co-crystallized with, the ligand compound. The resolution limit of the X-ray data has a great impact on the level of detail that will be revealed by the ED maps.

Figure 13.2 shows that, at a resolution 1.2 Å or better, individual atoms can be distinguished in the map, thus providing often an exquisite amount of detail for a ligand (such as indicating its exact chemistry). At around 2.0 Å resolution the map is less detailed but ligand placement will still be good with the data generally determining torsion angles well and often revealing details about ring pucker. At 3.0 Å resolution or worse, much less detail is available. Ligands can still normally be positioned with confidence, but it becomes increasingly essential to have prior knowledge of the chemistry of the ligand as the resolution worsens. However, at

Fig. 13.2 The effect of X-ray data resolution limit on the level of detail available. BUSTER-REPORT images of ED maps for the sucrose ligand in structures (**a**) 1YLT 1.2 Å resolution, (**b**) 2PWE 2.0 Å resolution and (**c**) 2QQV 3.0 Å resolution. In all three cases the placement and refinement of the sucrose ligand is good

low resolution many details are not available, and it must be borne in mind that features such as ring pucker may eventually be set as a consequence of the restraint dictionary and fitting procedures used, rather than on the basis of the X-ray data.

13.1.4 Data Collection Problems

The importance of collecting data correctly cannot be overstated. An example of a PDB entry where poor data collection directly affects the result is 1T0O [17]. BUSTER reports that the data are incomplete (Fig. 13.3a), and further analysis with the CCP4 program HKLVIEW shows that little data has been collected along the k axis (Fig. 13.3b). This results in a map with artifacts along the y-axis, causing the ED for the ligand to join up with that of the protein (Fig. 13.3c). Although the nominal data resolution limit of 1T0O is 1.96 Å [17], this systematic data incompleteness makes interpretation difficult. The only way to tackle data collection problems is to collect more and/or better data in the course of the experiment itself.

Global Phasing is currently helping a number of synchrotrons to provide users with strategies to collect better data for a given crystal.

13.1.5 Data Processing Problems

Once the X-ray data are collected, the resulting diffraction images must be processed and the Bragg diffraction spots integrated. There are a number of programs to do this and the topic is outside the scope of this presentation except to note that data processing must be correctly done to obtain meaningful results.

Many mistakes can be made at the data integration and other stages during the processing. A common error is to not properly tackle "ice rings" in the diffraction images [18, 19]. These are caused by the build-up of ice microcrystals on the protein

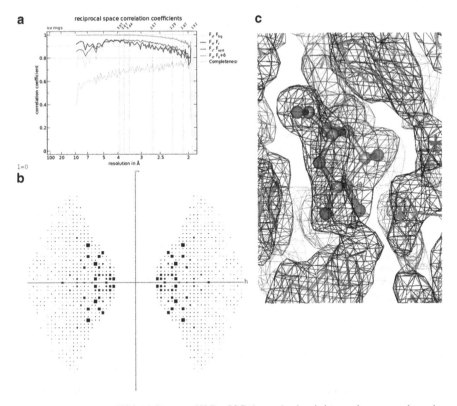

Fig. 13.3 PDB entry 1T0O (**a**) BUSTER [9] RecSSC shows the data is incomplete across the entire resolution range (**b**) HKLVIEW shows incomplete data is because no data were collected along k axis (**c**) BUSTER map after refinement shows density is poor along y direction, with the $2F_o$-F_c ED for the galactose ligand (*ball* and *stick*) merging with that for protein side-chains at *top* and *bottom*. This merging along the y-axis happens throughout the structure (the water molecule on the *left* provides another example)

crystal during data collection, and result in rings at characteristic resolutions. The affected resolution ranges should be excluded from all processing steps. Failure to do so has a detrimental effect on the internal scaling of the data, resulting in poor refinement and in ED map artifacts.

13.1.6 Is There Electron Density for the Ligand?

Given successful data processing, molecular replacement and initial refinement of the ligand-free protein model, the next step will be to assess the resulting ED maps for the presence of bound ligand, either at a known binding site or elsewhere. Pozharski et al. [20] emphasize that an unfortunately very common error is to believe that, because a ligand compound has been soaked, it must necessarily bind,

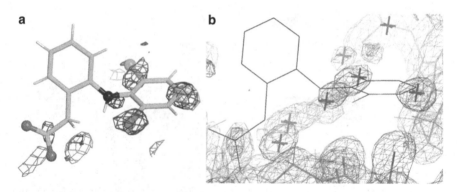

Fig. 13.4 Pozharski et al. [20] classify the diclofenac ligand in PDB entry 3IB0 [21] (1.4 Å resolution) as "absent". BUSTER-REPORT supports this classification: the ligand has high B-factors and a CC with the $(2F_o\text{-}F_c)$ map of 0.57, and as shown in panel (**a**) there is only a small amount of disconnected ED around it. Panel (**b**) shows a Coot image of the result of BUSTER re-refinement of the protein after the diclofenac (shown here as a *thin* "ghost") has been removed. The re-refinement included automated water placement, and shows that the ED can be well modeled by three water molecules (*crosses*) that form good hydrogen bonds (Color figure online)

and to model the ligand despite there being no evidence of its presence in the ED maps. For example, Pozharski et al. [20] classify the diclofenac ligand in PDB entry 3IB0 [21] as "absent". The BUSTER [9] map supports this classification (Fig. 13.4a). Revising the model by removing the diclofenac and refining with BUSTER including automated water molecule placement shows the ED into which the ligand had been placed can be well modeled by three water molecules (Fig. 13.4b).

By contrast, it is also possible to misinterpret ligand density as bound solvent. An interesting example of this is provided by PDB entry 2GWX as discussed in a review by Davis et al. [22]. In the original structure, ED in the ligand binding site was interpreted as being due to bound water molecules. Re-evaluation of the structure using the original SF by Fyffe et al. [23] led to the conclusion that this ED was actually due to a fatty acid ligand. In addition, clear density was found for n-heptyl-b-D-glucopyranoside (an additive in crystallization) in four sites. The revised structure is available as PDB entry 2BAW.

13.1.7 Producing the Restraint Dictionary for the Ligand

Given evidence in electron density to place the ligand, the next step is to fit a model into that density. Before this, it is necessary to produce an initial molecular model for the ligand, together with a restraint dictionary comprising a complete set of ideal bond distances and bond angles as well as listing chiral atoms and planar groups. Such a dictionary describes, typically using a CIF format, the chemical nature of the ligand, its molecular connectivity and its flexibility. That information is required not only to define the degrees of freedom available in fitting the ligand into its target ED and for manipulating ligands in COOT, but also to provide additional stereochemical information to packages such as REFMAC [16], BUSTER and phenix.refine [24] to maintain good molecular geometry during structure refinement in spite of the limited resolution of the X-ray data. Molecular mechanics force fields provide an alternative to simple restraint dictionaries [30]. BUSTER has recently been extended to allow the use of the MMFF94s force field for ligands (and force field conformational strain energy may provide and additional ligand validation metric).

GRADE [13] is the Global Phasing restraint dictionary generator. It takes a SMILES string or "mol2" file containing 3D coordinates of all atoms as input. Like BUSTER-REPORT, GRADE uses the CSD structures as the primary source of restraint information by invoking the MOGUL [11] program. MOGUL will rapidly analyze bond lengths, bond angles and many dihedral angles by finding CSD structures that contain similar chemical groups. Where MOGUL cannot provide information quantum chemical procedures are invoked. As well as being distributed with the BUSTER package GRADE can be used through the Grade Web Server [25].

A mistake in describing the stereochemistry of the ligand can result in the wrong ligand being fit and refined. Chiral inversions in carbohydrates are a good example. Smart et al. [26] describe how re-refinement of PDB entry 1DET using BUSTER and a GRADE dictionary corrected a chiral inversion in the ribose ring of the ligand: the re-refined model has been deposited as PDB entry 3SYU. Liebeschuetz et al. [27] mention PDB entry 2EVS [28] as a similar example, where the hexyl-beta-D-glucoside ligand has been refined and deposited with a chiral inversion of the anomeric carbon atom. This inversion can be corrected by re-refinement, but re-deposition has not yet been performed. An additional example is described in Sect. 13.2.1 where re-refinement is used to correct an inverted chiral atom in trehalose-6-phosphate in 1byk.

Figure 13.1 shows how BUSTER re-refinement of PDB entry 2H7P [29] with a GRADE [13] dictionary for the ligand markedly improves its fit to the ED. As shown in the first workshop practical given at the Erice School (now available online [2]) re-refinement also deals with stereochemistry issues raised by MOGUL. Most notably it alters the pucker of the 5-membered lactam ring and cyclohexyl rings to conformations seen in the CSD (Fig. 13.1d). A corrected structure 4TZT that has both a good fit to ED and ligand geometry has now been deposited into the PDB to replace 2H7P.

13.1.8 Ligand Fitting

Given suitable difference density and a restraint dictionary for the ligand, the next step is to exploit the flexibility of the ligand, as implicitly defined by that dictionary, to fit it into ED (either difference density, or 2Fo-Fc density). This can be done by hand using the COOT program, or by entrusting the task to an automated ligand fitter such as Global Phasing's RHOFIT or OpenEye's AFITT [30].

Ligand fitting becomes increasingly difficult as the data resolution limit worsens, because the ED will necessarily cease to reflect aspects of the ligand shape that have a decisive role in the selection of the correct ligand pose. An extreme example is the location of an extra copy of the 12-residue cyclic peptide in PDB entry 1OSG by Smart et al. [26] where knowledge of the conformation of the peptide was essential to be able to interpret the difference ED. The re-interpreted model including that extra copy of the ligand is available as PDB entry 3V56.

The importance of achieving a good ligand fit for structure-guided drug discovery is illustrated by the example of the inhibitor DDR1-IN-1 bound to DDR1 kinase domain by Kim et al. [31]. In the original published structure [28] and the associated PDB deposition 4BKI, the indolin-2-one ring of the inhibitor was positioned according to the inhibitor design so as to form two hydrogen bonds to the protein. BUSTER re-refinement of the structure with GRADE restraints [13] and evaluation of the ligand geometry with BUSTER-REPORT revealed to us that this ring positioning resulted in geometrical strain as well as in a strengthening of the difference ED, indicating that the ring should be flipped (Fig. 13.5a). The ligand placement after a ring flip and re-refinement is significantly better, with a good fit to the 2Fo-Fc

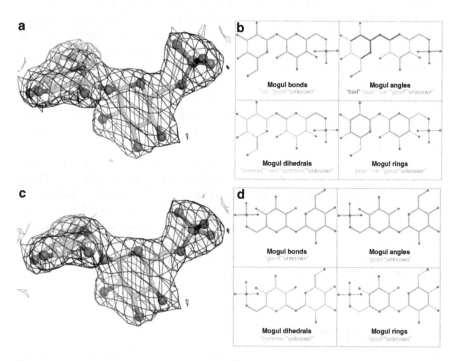

Fig. 13.5 (**a**) BUSTER re-refinement of 1BYK with a restraint dictionary for trehalose-6-phosphate specifying an incorrect anomer results in a structure with strong difference ED next to the inverted C1 atom. Panel (**b**) shows MOGUL results, indicating that refining with the inverted C1 atom produces a conformation with poor geometry. Re-refinement with a corrected trehalose-6-phosphate yields a much better fit to density (**c**) and alters MOGUL metrics to "good" or "common" (**d**)

density (Fig. 13.5b). After seeing this analysis, Canning, Bullock and co-workers deposited a corrected structure 4CKR and published a correction [32]. Given that the indolin-2-one ring fails to form the anticipated hydrogen bond contact and instead packs with the hydrophobic side of the ring adjacent to the main chain carbonyl of residue 702, there is a clear scope for revising the initial approach to designing a ligand that would form optimal interactions with the protein at that site.

13.2 Results

13.2.1 Achieving Correct Ligand Geometry in Trehalose Receptor Structure

The re-refinement of 1BYK provides an informative example of how a wrong assignment of chirality in a ligand can produce clear knock-on effects that are sensed by the metrics for both the ED fit and the ligand geometry. The structure is that of the *E. coli* Trehalose Receptor in complex with trehalose-6-phosphate, solved in 1998 at 2.5 Å resolution structure by Hars et al. [33]. Trehalose is a natural alpha-linked disaccharide formed by an α,α-1,1-glycosidic bond between two α-glucose units. However, the original 1BYK deposition used β-glucopyranose instead of the α-anomer for one of the sugar rings. This error propagated to the PDB chemical components dictionary [34, 35], giving rise to a definition of trehalose-6-phosphate T6P that specified the incorrect anomer, whereas the entry for trehalose itself was correct. BUSTER re-refinement of the structure with a GRADE [13] restraints dictionary for that incorrect anomer results in a structure with strong difference ED next to the inverted atom (Fig. 13.5a). Furthermore, MOGUL geometry validation through BUSTER-REPORT (Fig. 13.5b and Table 13.1) show that the geometry of the molecule is forced to be "unusual" because of the strain induced by fitting to ED that is not compatible with the model density for the ligand with its incorrect geometry.

Kay Diederichs and colleagues at the University of Konstanz asked for our assistance in correcting the structure. The raw diffraction data were re-processed with the current version of XDS [36] resulting in a dataset that had a completeness of 99.4 % compared to 67.5 % for the original. This demonstrates the importance of the retention of diffraction images [7]. Care was taken to ensure that the set of reflections used for R_{free} [37] was kept consistent with original structure factors. The next task was to produce a GRADE [13] restraints dictionary for T6P with the correct chirality. BUSTER re-refinement of the structure using this dictionary flipped the incorrect chiral centre without any further intervention. Following this re-refinement, the trehalose-6-phosphate fits the ED well with no difference density (Fig. 13.5c). In addition, all MOGUL metrics are altered to "good", showing that the trehalose-6-phosphate stereochemistry is now in complete agreement with that expected from related saccharides in the CSD (Table 13.1 and Fig. 13.5d). The model was improved by rounds of rebuilding using Coot and MolProbity [10] to assess geometry and ED fit. Table 13.1 shows how modern tools can achieve a

Table 13.1 Re-refinement of 1BYK correcting ligand geometry

	1byk.pdb	1byk.pdb re-refined[b] using T6P dictionary with incorrect trehalose	1byk.pdb re-refined[b] using corrected T6P dictionary	After multiple rounds of re-building and re-refinement[b]
BUSTER R_{work}	0.1935	0.1617	0.1604	0.1510
BUSTER R_{free}	0.1976	0.1871	0.1866	0.1730
$100*(R_{free} - R_{work})$	0.4 %	2.5 %	2.6 %	2.2 %
T6P ED fit CC $2F_o$-F_c[a]	0.961	0.978	0.985	0.988
T6P Mogul "bad" angles (#$|Z| > 4$)[a]	6	4	0	0
T6P Mogul "unusual" dihedrals/rings[a]	2/0	1/1	0/0	0/0
Number of water molecules placed	44	44	44	79
MolProbity Ramachandran outliers	0.8 %	0.4 %	0.2 %	0 %
MolProbity Ramachandran favored	94.9 %	97.0 %	96.8 %	98.4 %
MolProbity side chains with poor rotamers	6.6 %	7.1 %	7.4 %	1.3 %
MolProbity Overall Score/Percentile	2.13/92nd	1.64/99th	1.56/99th	0.73/100th

[a]Figure given for A chain copy only and the B chain values are similar
[b]BUSTER –autoncs option used [26]

structure that has improved interpretation and much better "quality metrics" than in 1998. This is a good example of the process of the mutual improvement of X-ray crystallographic software and structure models in the PDB [7]. It should be noted that the conclusions drawn by Hars et al. [33] from the original structure are unaffected. The corrected structure has been deposited in the PDB and will obsolete the original 1BYK entry. The PDB chemical components dictionary [34, 35] definition of trehalose-6-phosphate T6P has also been updated.

13.2.2 New Insights into the Ligand Geometry in a JNK3 Kinase Structure

The PDB entry 1PMQ for the JNK3 kinase complex [38] provides an interesting example of how advances in methodology can lead to improvements in the modeling of a ligand. Note that the material here forms the basis for the second workshop practical session given at the Erice School, now available online [2].

1PMQ is the structure of JNK3 in complex with an imidazole-pyrimidine inhibitor, solved in 2003 by Giovanna Scapin and colleagues at Merck [38]. The ligand has been assigned the three-letter code 880 in the PDB chemical components dictionary [34, 35]. Visual inspection of the deposited PDB entry together with ED maps from BUSTER shows that the model for ligand 880 fits the density well [2]. However, MOGUL analysis as provided by BUSTER-REPORT [8] shows that it would be expected from CSD structures that the atom C55 of the cyclohexyl ring should be coplanar with the pyrimidine ring in the ligand, but that this is not the case in 1PMQ. Simple re-refinement with BUSTER using a GRADE [13] restraints dictionary for 880 cannot fix the problem, but once the cyclohexyl ring is flipped manually re-refinement achieves a good fit to density, good ligand geometry and improved geometry for the protein-ligand hydrogen bonds [2].

As the data set has a high degree of anisotropy the Diffraction Anisotropy Server [39] was used, producing a noticeable improvement in map quality. Inspection of the ED enables further improvements to the structure. The dichlorophenyl ring in the 880 ligand shows difference density near atom CL45, indicating that the ring has two alternate conformations that can be modelled [2]. The improved maps and model support the identification by Scapin et al. [38] of the "accidental" second ligand AMP-PCP as well as a subtle improvement in its ED fit and geometry. After additional rounds of rebuilding the protein/solvent, the corrected structure [2] has now been deposited in the PDB obsoleting the original 1PMQ entry. Once again improvements to the structure are limited and only add support to the conclusions drawn by Scapin et al. [38].

13.3 Discussion

Many researchers have pointed out that there are problems in the chemistry, placement and fit of ligands in the PDB [39, 30, 40–42, 27, 43, 44] and have asked what can be done to improve matters. The implementation of the recommendations of wwPDB X-ray crystallographic task force [45], and in particular the use of MOGUL analysis as part of the deposition process, will hopefully contribute to avoiding problems in current and future depositions. A critical factor in this context is "the urgent need to provide adequate training to next-generation crystallographers" as noted by Dauter et al. [46]. It is hoped that this chapter, together with the accompanying workshop practical material [2], will make a positive contribution, however small it may be, towards this goal.

Improving matters for the future is essential, but what can be done about problems with existing PDB entries? One solution is to produce secondary databases containing re-refined, corrected and/or curated structures, as has been done as part of the PDB-REDO [47] and IRIDIUM [43] projects. This is a valuable approach but its usefulness is likely to be restricted to a small number of specialist users. It is the PDB [1] itself that is the vital resource for many non-specialists, and it is a regrettable that problematic entries very often persist in the database.

On reflection, what is unacceptable is to criticize PDB entries to the extent of proposing alternative models, without taking any corrective action nor being required to do so. In many respects this is fundamentally unfair to both the original depositors and the users of the PDB. Journals now all require deposition into the PDB of any structures reported in a paper. Although this criterion is universally applied to the reports of *new* structures, it appears not to be applied to publications pointing out errors in *existing* (i.e. already deposited) structures, even when an alternative, purportedly improved structure is shown in a figure. Matters are made worse by the fact that the allegedly problematic structures are only referred by their PDB code, without citing the original reference. The practical upshot of this is that a researcher relying on such a PDB entry has little chance of ever becoming aware that it has been called into question in a published paper, as a literature survey would fail to find a reference to the latter. It would be particularly annoying for any researcher who used a PDB result to find out that a correction was available but had gone unrecorded.

On occasion, complete deposition may not be straightforward because the result originates from a molecular modeling method. In such cases, the best option in the first instance is to contact the authors of the deposition(s) and suggest that revision and re-deposition is in order. Failing this, it may be possible to find a friendly protein crystallographer to produce a re-refined corrected structure and deposit this as a "REMARK 0 alternative interpretation", including the methodology used as part of the publication. Such an entry is given a separate PDB code and does not replace the original entry in the PDB. If it is not possible to achieve a deposition in the actual PDB, then at the minimum the coordinates of the proposed alternative model should be included as Supplementary Material that will thus be available with the publication. It is important to include a citation of the original publication to ensure that users of a structure will more easily be able to find relevant information about its amended versions. Paper reviewers should encourage deposition whenever possible.

We hasten to add that in the past we have been guilty of exactly the behavior that we criticize above. However, our intention is to ensure that corrected entries appear in the PDB wherever possible. This should further invigorate the recently analyzed process of continuous mutual improvement of macromolecular structure models in the PDB and of X-ray crystallographic software [7].

To conclude, we would strongly recommend that users of ligand complexes from the PDB take a cautious approach and make full use of the critical assessment tools available at the time when they wish make use of an existing entry, however recently it may have been deposited, as those tools may themselves have improved.

Acknowledgements Our warmest thanks go to our colleagues at Global Phasing for discussions, provision of examples and help with these notes. This work was supported by members of the Global Phasing Consortium and by the SILVER FP7-HEALTH Integrated Project of the European Commission under grant agreement no. 260644. We thank all the crystallographers who deposited into the PDB the structures that have been analyzed here. Special thanks to Kay Diederichs, Giovanna Scapin and colleagues. Critical assessment of the structures allows for improvements to be made in methods, so that problems can be corrected in existing structures and avoided in future structures.

References

1. Berman HM, Westbrook J, Feng Z, Gilliland G, Bhat TN, Weissig H, Shindyalov IN, Bourne PE (2000) The protein data bank. Nucleic Acids Res 28(1):235–242
2. Smart OS (2014) Achieving high quality ligand chemistry in protein X-ray structures. Workshop prepared for the "Structural Basis of Pharmacology: Deeper Understanding of Drug Discovery through Crystallography", Meeting Erice June 2014. http://grade.globalphasing.org/tut/erice_workshop/
3. Leach AR, Gillet VJ (2003) An introduction to chemoinformatics. Kluwer Academic Publishers, Dordrecht/Boston
4. Nissink JWM, Murray C, Hartshorn M, Verdonk ML, Cole JC, Taylor R (2002) A new test set for validating predictions of protein-ligand interaction. Proteins 49(4):457–471
5. Morris GM, Huey R, Lindstrom W, Sanner MF, Belew RK, Goodsell DS, Olson AJ (2009) AutoDock4 and AutoDockTools4: automated docking with selective receptor flexibility. J Comput Chem 30(16):2785–2791
6. Sharff A, Keller P, Vonrhein C, Smart O, Womack T, Flensburg C, Paciorek W, Bricogne G (2014) Pipedream documentation, version 1.0.0. Global Phasing Ltd. http://www.globalphasing.com/buster/manual/pipedream/manual/index.html
7. Terwilliger TC, Bricogne G (2014) Continuous mutual improvement of macromolecular structure models in the PDB and of X-ray crystallographic software: the dual role of deposited experimental data. Acta Crystallogr D Biol Crystallogr 70(Pt 10):2533–2543
8. Buster-report. Global Phasing Ltd. https://www.globalphasing.com/buster/wiki/index.cgi?BusterReport
9. Bricogne G, Blanc E, Brandl M, Flensburg C, Keller P, Paciorek W, Roversi P, Sharff A, Smart OS, Vonrhein C, Womack TO (2014) BUSTER version 2.13.0. Global Phasing Ltd., Cambridge, UK
10. Chen VB, Arendall WB, Headd JJ, Keedy DA, Immormino RM, Kapral GJ, Murray LW, Richardson JS, Richardson DC (2010) MolProbity: all-atom structure validation for macromolecular crystallography. Acta Crystallogr Sect D: Biol Crystallogr 66:12–21
11. Bruno IJ, Cole JC, Kessler M, Luo J, Motherwell WDS, Purkis LH, Smith BR, Taylor R, Cooper RI, Harris SE, Orpen AG (2004) Retrieval of crystallographically-derived molecular geometry information. J Chem Inf Comput Sci 44(6):2133–2144
12. Allen FH (2002) The Cambridge Structural Database: a quarter of a million crystal structures and rising. Acta Crystallogr Sect B Struct Sci 58:380–388
13. Smart OS, Holstein J, Womack T (2014) Grade documentation. version 1.2.8. http://www.globalphasing.com/buster/manual/grade/manual/index.html
14. Emsley P, Lohkamp B, Scott WG, Cowtan K (2010) Features and development of Coot. Acta Crystallogr Sect D: Biol Crystallogr 66:486–501
15. Kleywegt GJ, Harris MR, Zou JY, Taylor TC, Wahlby A, Jones TA (2004) The Uppsala electron-density server. Acta Crystallogr Sect D: Biol Crystallogr 60:2240–2249
16. Murshudov GN, Skubak P, Lebedev AA, Pannu NS, Steiner RA, Nicholls RA, Winn MD, Long F, Vagin AA (2011) REFMAC5 for the refinement of macromolecular crystal structures. Acta Crystallogr Sect D: Biol Crystallogr 67:355–367
17. Golubev AM, Nagem RAP, Neto JRB, Neustroev KN, Eneyskaya EV, Kulminskaya AA, Shabalin KA, Savel'ev AN, Polikarpov I (2004) Crystal structure of alpha-galactosidase from Trichoderma reesei and its complex with galactose: implications for catalytic mechanism. J Mol Biol 339(2):413–422
18. Rupp B (2010) Biomolecular crystallography: principles, practice, and application to structural biology. Garland Science, New York
19. Vonrhein C, Flensburg C, Keller P, Sharff A, Smart O, Paciorek W, Womack T, Bricogne G (2011) Data processing and analysis with the autoPROC toolbox. Acta Crystallogr Sect D: Biol Crystallogr 67:293–302

20. Pozharski E, Weichenberger CX, Rupp B (2013) Techniques, tools and best practices for ligand electron-density analysis and results from their application to deposited crystal structures. Acta Crystallogr Sect D: Biol Crystallogr 69:150–167

21. Mir R, Singh N, Vikram G, Kumar RP, Sinha M, Bhushan A, Kaur P, Srinivasan A, Sharma S, Singh TP (2009) The structural basis for the prevention of nonsteroidal antiinflammatory drug-induced gastrointestinal tract damage by the C-lobe of bovine colostrum lactoferrin. Biophys J 97(12):3178–3186. doi:10.1016/j.bpj.2009.09.030

22. Davis AM, St-Gallay SA, Kleywegt GJ (2008) Limitations and lessons in the use of X-ray structural information in drug design. Drug Discov Today 13(19–20):831–841. doi:10.1016/j.drudis.2008.06.006

23. Fyffe SA, Alphey MS, Buetow L, Smith TK, Ferguson MAJ, Sorensen MD, Bjorkling F, Hunter WN (2006) Reevaluation of the PPAR-beta/delta ligand binding domain model reveals why it exhibits the activated form. Mol Cell 21(1):1–2

24. Afonine PV, Grosse-Kunstleve RW, Echols N, Headd JJ, Moriarty NW, Mustyakimov M, Terwilliger TC, Urzhumtsev A, Zwart PH, Adams PD (2012) Towards automated crystallographic structure refinement with phenix.refine. Acta Crystallogr Sect D: Biol Crystallogr 68:352–367

25. Grade Web Server. Global Phasing Ltd. http://grade.globalphasing.org/

26. Smart OS, Womack TO, Flensburg C, Keller P, Paciorek W, Sharff A, Vonrhein C, Bricogne G (2012) Exploiting structure similarity in refinement: automated NCS and target-structure restraints in BUSTER. Acta Crystallogr Sect D: Biol Crystallogr 68:368–380

27. Liebeschuetz J, Hennemann J, Olsson T, Groom CR (2012) The good, the bad and the twisted: a survey of ligand geometry in protein crystal structures. J Comput Aided Mol Des 26(2):169–183

28. Malinina L, Malakhova ML, Kanack AT, Lu M, Abagyan R, Brown RE, Patel DJ (2006) The liganding of glycolipid transfer protein is controlled by glycolipid acyl structure. PLoS Biol 4(11):1996–2011

29. He X, Alian A, Stroud R, de Montellano PRO (2006) Pyrrolidine carboxamides as a novel class of inhibitors of enoyl acyl carrier protein reductase from Mycobacterium tuberculosis. J Med Chem 49(21):6308–6323

30. Wlodek S, Skillman AG, Nicholls A (2006) Automated ligand placement and refinement with a combined force field and shape potential. Acta Crystallogr Sect D: Biol Crystallogr 62:741–749

31. Kim HG, Tan L, Weisberg EL, Liu FY, Canning P, Choi HG, Ezell SA, Wu H, Zhao Z, Wang JH, Mandinova A, Griffin JD, Bullock AN, Liu QS, Lee SW, Gray NS (2013) Discovery of a potent and selective DDR1 receptor tyrosine kinase inhibitor. ACS Chem Biol 8(10):2145–2150

32. Kim HG, Tan L, Weisberg EL, Liu FY, Canning P, Choi HG, Ezell S, Zhao Z, Wu H, Wang JH, Mandinova A, Bullock AN, Liu QS, Lee SW, Gray NS (2014) Discovery of a potent and selective DDR1 receptor tyrosine kinase inhibitor (vol 8, pg 2145, 2013). ACS Chem Biol 9(3):840

33. Hars U, Horlacher R, Boos W, Welte W, Diederichs K (1998) Crystal structure of the effector-binding domain of the trehalose-repressor of Escherichia coli, a member of the LacI family, in its complexes with inducer trekalose-6-phosphate and noninducer trehalose. Protein Sci 7(12):2511–2521

34. Feng ZK, Chen L, Maddula H, Akcan O, Oughtred R, Berman HM, Westbrook J (2004) Ligand Depot: a data warehouse for ligands bound to macromolecules. Bioinformatics 20(13):2153–2155

35. Dimitropoulos D, Ionides J, Henrick K (2006) Using MSDchem to search the PDB ligand dictionary. Current protocols in bioinformatics/editorial board, Andreas D Baxevanis [et al] Chapter 14:Unit14.13

36. Kabsch W (2010) XDS. Acta Crystallogr Sect D: Biol Crystallogr 66:125–132

37. Brunger AT (1992) Free R-value – a novel statistical quantity for assessing the accuracy of crystal-structures. Nature 355(6359):472–475

38. Scapin G, Patel SB, Lisnock J, Becker JW, LoGrasso PV (2003) The structure of JNK3 in complex with small molecule inhibitors: structural basis for potency and selectivity. Chem Biol 10(8):705–712
39. Davis AM, Teague SJ, Kleywegt GJ (2003) Application and limitations of X-ray crystallographic data in structure-based ligand and drug design. Angew Chem Int Ed 42(24):2718–2736
40. Kleywegt GJ (2007) Crystallographic refinement of ligand complexes. Acta Crystallogr Sect D: Biol Crystallogr 63:94–100
41. Joosten RP, Womack T, Vriend G, Bricogne G (2009) Re-refinement from deposited X-ray data can deliver improved models for most PDB entries. Acta Crystallogr Sect D: Biol Crystallogr 65:176–185
42. Malde AK, Mark AE (2011) Challenges in the determination of the binding modes of nonstandard ligands in X-ray crystal complexes. J Comput Aided Mol Des 25(1):1–12
43. Warren GL, Do TD, Kelley BP, Nicholls A, Warren SD (2012) Essential considerations for using protein-ligand structures in drug discovery. Drug Discov Today 17(23–24):1270–1281
44. Reynolds CH (2014) Protein-ligand cocrystal structures: we can do better. ACS Med Chem Lett 5(7):727–729
45. Read RJ, Adams PD, Arendall WB, Brunger AT, Emsley P, Joosten RP, Kleywegt GJ, Krissinel EB, Lutteke T, Otwinowski Z, Perrakis A, Richardson JS, Sheffler WH, Smith JL, Tickle IJ, Vriend G, Zwart PH (2011) A new generation of crystallographic validation tools for the protein data bank. Structure 19(10):1395–1412
46. Dauter Z, Wlodawer A, Minor W, Jaskolski M, Rupp B (2014) Avoidable errors in deposited macromolecular structures: an impediment to efficient data mining. Int Union Crystallogr J 1(Pt 3):179–193
47. Joosten RP, Joosten K, Murshudov GN, Perrakis A (2012) PDB_REDO: constructive validation, more than just looking for errors. Acta Crystallogr Sect D: Biol Crystallogr 68:484–496

Chapter 14
Molecular Obesity, Potency and Other Addictions in Drug Discovery

Michael M. Hann

Abstract Achieving the right balance of properties in a candidate drug molecule is a very complex challenge as many of them are in conflict with each other. Structure Based Drug Design is a key tool in the medicinal chemists toolkit but can lead to an over dependence on potency if not used in conjunction with physical chemistry predictions and measurements to maintain the property balance needed.

14.1 Introduction

The title of this chapter is taken from our 2011 publication, with the same title as this chapter, and this should be read in conjunction with this chapter for further background [1].

Drug discovery is a very complex activity that is often said to make rocket science look easy! Figure 14.1 attempts to summarize the journey that is required in both a multidimensional and multi-objective sense to attain the sweet spot where all the properties required of a safe and efficacious new medicine are appropriately balanced. Of course the view of the challenge of drug discovery presented in Fig. 14.1 can be over simplistic when we consider that there may not actually be a compromise that can be found between these conflicting properties. This may be because the target protein may actually be undruggable with a small molecule, or the window of specificity is vanishingly small.

A consequence of the complexity of our challenge is the balance between the genuinely predictable scientific activities ("which needs maths") from the more chaotic activities ("which need experience and intuition"). Another way of thinking about this is embedded in the truism that "the interesting things in science are the differences between theory and experiment"!

Protein crystallography and Structure-Based Drug Design (SBDD) have become key components of our toolkit to aid us on the journey, however using them without

M.M. Hann (✉)
GlaxoSmithKline Research and Development, Chemical Sciences, Molecular Discovery Research, Gunnels Wood Rd., Stevenage SG1 2NY, UK
e-mail: mike.m.hann@gsk.com

© Springer Science+Business Media Dordrecht 2015
G. Scapin et al. (eds.), *Multifaceted Roles of Crystallography in Modern Drug Discovery*, NATO Science for Peace and Security Series A: Chemistry and Biology, DOI 10.1007/978-94-017-9719-1_14

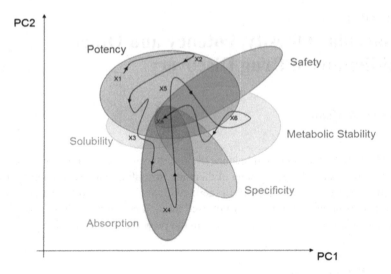

Fig. 14.1 A simplified 2D map of the challenge of drug discovery and the journey a project might take in starting from x1 and ending at xn

considering the bigger picture can lead to unfortunate consequences through the design of inappropriate molecules! This chapter aims to put in context some of the conflicts that a quality molecule needs to have in order to be a successful medicine and highlights the danger of seeking quick fixes through potency based on lipophilicity and other physiochemical influences.

While creativity in medicinal chemistry is at the heart of the drug discovery process it is often disciplines such as computational chemistry and structural biology that enable some of this creativity. The combined skill sets that are required for the identification of the best leads and then nurturing them through lead optimization on a complex landscape of constraints is often the defining characteristic of successful drug discovery campaigns.

14.2 What Are the Leading Causes of Failure in Drug Discovery?

It is clear from data compiled by Kola and Landis that comparison of cited reasons for drug discovery failures (e.g., attrition) in 1991 to those in 2000 showed a shift to an increase in failure attributed to toxicological reasons [2].

The issues of PK and bioavailability that were the leading cause of attrition in the 1991 data appear to have been largely controlled. It seems likely that this is due to both a better understanding of pharmacokinetic issues and also in the improved formulation of compounds that results in more chemical entities overcoming this hurdle. Some of this improvement in getting compounds into the body (and keeping them there) is likely related to the rise in attrition due to toxicological issues as the body responds to chemical entities that are "forced" into the body. Thus by using formulation technologies to deliver inappropriate molecules we may have only delayed failure (i.e., the realization that a molecular series is inappropriate) to a more expensive part of the drug discovery activity.

So what defines inappropriate molecules? Clearly some of the toxicity of a molecule or series will be due to activity at the intended intervention target (or its pathway), which may only become apparent during development. However it is becoming increasingly clear that there are more generic influences behind a significant proportion of toxicology-related attrition. Much of this realization has come from recently published analyses of internal data from large pharma companies and this has led to the emergence of new guidelines aimed at hopefully controlling this aspect of attrition in the future (e.g., [3]).

The earliest of the rules of thumb that have become prevalent in contemporary drug discovery parlance is the Lipinski "rule of fives" which has been adopted, and often erroneously used, as defining the limits of drug like space [4]. Lipinski's rule actually refers to the likelihood of a compound having oral bioavailability, based on a set of compounds that made it to Phase IIa and were therefore assumed to be a good indicator of oral absorption. Thus compounds which have one or more of either CLogP greater than 5, Molecular Weight greater than 500, Number of H-bond acceptors greater than 10 or Number of H-bond donors greater than 5 are less likely to be orally absorbed. It is now becoming increasingly clear that a much more tightly defined set of rules are appropriate if we are considering drug space from the viewpoint of toxicological risk rather than the risk of a compound not being orally absorbed.

The publication from Leeson and Springthorpe at AstraZeneca on Receptor Promiscuity clearly highlighted (Fig. 14.2) the problem of excessive lipophilicity and they introduced the term Lipophilic Ligand Efficiency (LLE) (defined as pIC50 – cLogP) to help highlight likely promiscuous compounds [3]. If LLE >5, then the compound related toxicity risk is greatly reduced. The reason for this can be readily understood when it is remembered that the lipophilicity scales (such as cLogP) are logarithmic and therefore an increase of just one unit in cLogP means that there is now ten times more compound present in the highly lipophilic cellular membranes. These membranes are the home to many of the critical signaling systems and inappropriate triggering by local high concentrations of not very intrinsically potent compounds, can easily lead to unwanted effects leading to potentially toxicological events. While promiscuity of this non-specific type will be detrimental, there may be situations (e.g. for polypharmacology) where some degree of promiscuity is of course desirable.

In another study Hughes et al. at Pfizer showed (Fig. 14.3) that compounds with a cLogP <3 & Total Polar Surface Area TPSA >75 have a sixfold reduced in vivo toxicity compared to cLogP >3 and TPSA <75. This is known as the Pfizer 3/75 rule [5].

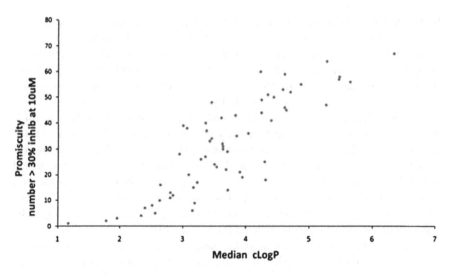

Fig. 14.2 Rise in promiscuity of compounds with cLogP (Data from Ref. [3])

Probabilities of toxicity vs ClogP/TPSA				
	based on [total drug]		based on [free drug]	
	TPSA>75	TPSA<75	TPSA>75	TPSA<75
ClogP<3	0.4	1.1	0.4	0.5
ClogP>3	0.4	2.4	0.8	2.6

Fig. 14.3 Toxicity probability matrix based on TPSA and cLogP criteria (Data from Ref. [5])

Another example (Fig. 14.4) of an analysis that has led to the emergence of a further set of guidelines is the GSK 4/400 rule which relates to compounds with a CLogP less than 4 and a MW less than 400 and how on average they have a more favorable ADMET profile [6]. This analysis by Paul Gleeson looked at ca. 30,000 neutral, basic, acidic and zwitterionic molecules that have been profiled in multiple physical chemistry and ADMET assays at GSK.

A further perspective on attrition related parameters is given by a count of the number of aromatic rings. While clearly related in a non-linear manner to lipophilicity, the analysis of this property by Ritchie and MacDonald gives interesting insights which can be summarized as "the fewer aromatic rings in an oral drug candidate the better" with less than three being suggested as an appropriate target number [7]. This use of a count of aromatic rings has been extended (Fig. 14.5) by Young et al. at GSK to define a Property Forecast Index PFI as the sum of a chromatographically measured logD at pH 7.4 and the number of aromatic rings. If PFI is <6 then compounds are likely more soluble and have reduced ADMET risks [8].

Continuing this theme of the dangers of too much sp2 or aromatic character in molecules, Humblet et al. showed that the survival rate of compounds through the

neutral molecules	MW<400 & cLogP<4		MW>400 & cLogP>4	
solubility	average	intermediate	lower	bad
permeability	higher	good	average/higher	intermediate
bioavailability	average	intermediate	lower	bad
volume of distribution	average	intermediate	average	intermediate
plasma protein binding	average	intermediate	higher	bad
CNS penetration	higher/average	bad	average/lower	good
brain tissue binding	lower	bad	higher	bad
P-gp efflux	average	intermediate	higher/average	bad
in vivo clearance	average	intermediate	average	intermediate
hERG inhibition	lower	good	lower	good
P450 inhibition	lower 2C9, 2C19, 2D6 & 3A4 inhib	good	higher 2C9, 2C19 & 3A4 inhib	bad
	higher 1A2 inhib	bad	lower 1A2 inhib	good
			average 2D6 inhib	intermediate

Fig. 14.4 Effect of GSK 400/4 rule on properties effecting ADMET properties (Data from Ref. [6]). Good, Intermediate or Bad refers to criteria for a peripheral target

	PFI = mChrom log D7.4 + #Ar								
Assay/target value	<3	3-4	4-5	5-6	6-7	7-8	8-9	9-10	>10
Solubility >200 µM	89	83	72	58	33	13	5	3	2
&HSA <95%	88	80	74	64	50	30	17	8	4
2C9 pIC50 <5	97	90	83	68	48	32	23	22	38
2C19 pIC50 <5	97	95	91	82	67	52	42	42	56
3A4 pIC50 <5	92	83	80	75	67	60	58	61	66
Clint <3ml/min/kg	79	76	68	61	54	42	41	39	52
Papp>200nm/s	20	30	46	65	74	77	65	50	33

Fig. 14.5 Property Forecast Index PFI and its effect on a number of ADMET properties. Numbers show percentage of compounds achieving defined target values in various developability assays (Data from Ref. [8])

drug discovery process is enhanced by an increase in the fraction of sp3 hybridized carbon atoms and the number of chiral centers present [9]. In addition to giving access to a greater diversity of compounds to explore it seems likely that one of the benefits of chirality in a drug is that it leads to increased complexity (and hence potential potency through appropriate complementarity) to a specific target without increasing the molecular weight of ligands [10].

Lipophilicity is well known to be the antithesis of solubility and lack of solubility has been a consistent problem for medicinal chemists [11]. As noted earlier improved formulation methodologies can somewhat mitigate this situation. However relying on formulation to get insoluble compounds on board is likely only to aggravate the body to work harder to eliminate them. The usual response of the body to lipophilic xenobiotics is to try to make them more polar via metabolism so they can be excreted. Medicinal chemists are then faced with the need to make their lipophilic and insoluble compounds more metabolically stable to prolong their duration of actions. This can put enormous demands on a compounds profile, especially if once a day dosing is being sought as the target product profile. Blocked metabolism means the body will need to find more extreme ways of removing the compound, often inducing more high potential and thus intrinsically more reactive and toxic species.

If as a result of our lead optimization we end up with excessively large and lipophilic molecules, it is clear that we have likely embedded other properties into these molecules that will limit their ability to become successful medicines. In effect they have become too large and too lipophilic for their own good and for this reason we introduced the term Molecular Obesity. As with medical obesity, which is measured by Body Mass Index (BMI), medicinal chemists have developed their own indices (Fig. 14.6 shows a summary of some of these) such as Ligand Efficiency LE (= binding affinity/number of heavy atoms) [12] and the already mentioned Lipophilic Ligand Efficiency LLE [3] to help identify and control the effects of molecular obesity, which are implicated in the premature demise of far too many drug candidates in recent years. Additional indices continue to emerge in the literature, all with the aim of restricting the tendency to Molecular Obesity. One such index is the LLE_{AT} index, which combines aspects of LE and LLE and adjusting so that the value is on the same scale as LE, so again 0.3 is a good target

---

- **Ligand Efficiency Index (LE)**
  Ligand efficiency: a useful metric for lead selection Hopkins et al.. DDT 2004;9(10):430-1
  LE=ΔG/#heavy-atoms
  Potency in kcal/mol (=-1.37logKd) normalised by dividing by the number of heavy atoms
  An *'idealised'* compound with 1nm pIC50 and 30 heavy atoms has LEI = 0.42
  An *'okay'* compound with 10nm pIC50 and 38 heavy atoms (MW 500) has LEI = 0.29

- **Lipophilic Ligand Efficiency Index (LLE)**
  The influence of drug-like concepts on decision-making in medicinal chemistry. Leeson and Springthorpe . NRDD 2007;6(11):881-90
  LLE = pIC50 – clogP
  Potency normalised by lipophilicity to ensure specific rather than non-specific effects.
  Typical good value are 5-7 for nanomolar potency compound

- **Ligand Ligand Efficiency (Astex version)**
  Assessing the lipophilicity of fragments and early hits. Mortenson and Murray JCAMD, 2011, 25 (7), 663-667
  LLE_{AT} . 0.11*ln(10)*RT(logP-log(Kd))/HA
  Combined LLE and LE index which is parameterised to be on same scale as LE, so 0.3 is considered a good target value. Particularly good for assessing fragment hits.

- **Binding Efficiency Index (BE)**
  Ligand efficiency indices as guideposts for drug discovery. Abad-Zapatero and Metz. Drug Discovery Today, 10(7), 2005, 464-469
  BE = pIC50/MW.   Potency (pIC50) normalised for MW
  An *'idealised'* compound with 1nm pIC50 and MW of 0.333 kDA has BEI = 27

- **Surface Binding Efficiency Index (SE)**
  Recent Developments in Fragment-Based Drug Discovery J. Med. Chem.; Congreve et al.. 2008; 10;51(13):3661-80;
  Potency normalised for Polar Surface Area
  An *'idealised'* compound with 1nm pIC50 and PSA of $50A^2$ has SEI = 18

Fig. 14.6 Summary of medicinal chemistry indices and guidance on target values

value [13]. This index is particularly useful for assessing early fragment hits for follow up.

14.3 Driving Potency Through Molecular Obesity

We like potency in our molecules for a number of reasons and it is worth examining why this is. This is a particular problem when following a SBDD approach where the availability of compelling structural insights enable by progress in protein crystallography can lead to highly potent molecules designed and built to fit the protein target. While there is nothing intrinsically wrong with potent molecules, there is when other properties compromise their overall effectiveness!

One of the basic tenets of medicinal chemistry is that increasing ligand potency leads to increased specificity and hence to an improved therapeutic index [14]. This is true if the potency is based around directional (i.e., polar) interactions because the directionality implies specificity. In 2001, we introduced the concept of Molecular Complexity (Fig. 14.7) [10] and updated this in a further paper in 2011 [15]. The basic tenet of the idea is embedded in a simple and abstract model of molecular interactions between a ligand and receptor that gives insights into the probability of finding appropriate complementarity at different levels of molecular complexity. While we aspire to find very complex and thus potentially potent (and specific) interactions the chances of finding these all at once (i.e., in HTS) is highest when we only expect to get a few right initially. This is the basis of the fragments approach, which is based on finding weakly binding but small compounds with just a minimal number of correct interactions. We then iteratively grow the molecule to find new interactions and hence potency. However one of the easiest ways of gaining potency is through lipophilic interactions which are non-directional and therefore do not require precise engineering. In a recent book chapter we have developed an extension of the complexity model, which uses information content as a way of representing (Fig. 14.8) such non-directional interactions [16]. In the left hand representation, slippage is difficult as the complexity in the pattern makes slippage difficult. While on the right low information content (e.g., lipophilicity) can slip easily and in addition all the secondary interactions as attractive.

Of course high potency can allow reduction in the size of dosage especially if a compound has good pharmacokinetic properties. Low dosage not only helps reduce the cost of goods but it is also one of the only known predictors of low incidence of idiosyncratic toxicity [28]. Potency can compensate for low bioavailability in that the small portion of, for example, a poorly absorbed drug that does get into the circulation will at least have a chance of being efficacious if it has high molar potency. However this brings a high risk in that the part of the high dosage that is not being effectively utilized is available to cause off target issues.

So while there are many reasons why potency is a good thing, the problem is how we have often gone about achieving it. Ladbury and colleagues have shown by ITC studies that the Free Energy of interaction (i.e. potency) of synthetic ligands

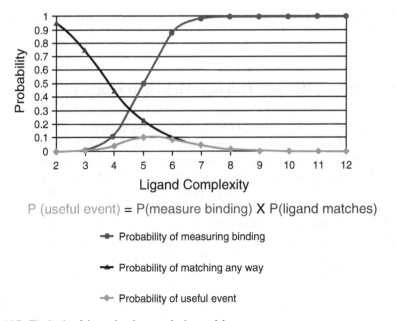

$$P \text{ (useful event)} = P(\text{measure binding}) \times P(\text{ligand matches})$$

➡ Probability of measuring binding

➡ Probability of matching any way

➡ Probability of useful event

Fig. 14.7 The basis of the molecular complexity model

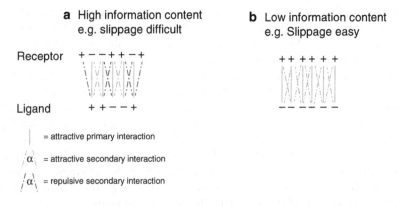

Fig. 14.8 Influence of information content on slippage ability in molecular complexity model

correlates well with the ligand's hydrated apolar surface area (i.e. lipophilicity) that is buried in the interaction [17]. This shows that we tend to use the easy gains of potency by adding lipophilicity.

Again using ITC data on a large number of compounds, Keseru et al. showed (Fig. 14.9) that as potency increases, enthalpic contributions tend to a maximum and then starts to fall while entropy starts to rise further in the most potent compounds [18, 19]. Broadly speaking enthalpy equates to polar interactions while a key contributor to entropy is lipophilic interactions. This suggests that if you do not

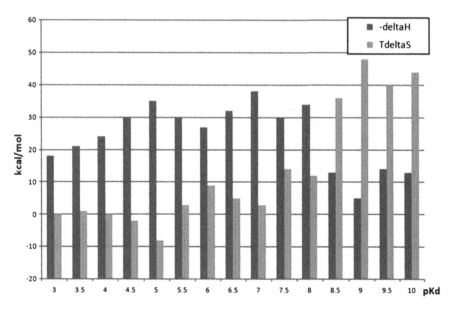

Fig. 14.9 Contribution of enthalpy (*light bars*) and entropy (*dark bars*) to overall potency of compounds (ordered by increasing potency) (Redrawn from original data used in Ref. [19])

get the maximum available enthalpic binding of a fragment or template to start with then you will end up having to use entropic interactions such as lipophilicity to get the desired potency. The use of indices such as LE and LLE are particularly useful for this purpose.

In other surveys of specific and cross-series data it is also apparent that increasing molecular weight and complexity tend to correlate with increased potency [20]. This is consistent with adding either specific or non-specific interactions but it is also likely to be a consequence of the predilection of medicinal chemists, who invariably trained as synthetic organic chemists, to build molecules rather than take them apart. Interestingly the equivalent of retro-synthetic analysis, which is so critical to planning good syntheses, has only recently become more embedded into the medicinal chemistry sphere in terms of fragmenting hits to find the most ligand efficient and smallest critical part [21]. An additional aspect of synthetic chemistry that has only recently been shown is how even laboratory practices such as reaction work up actually bias the synthesis of more lipophilic compounds, presumably as they are more easily extracted from the aqueous reaction quench [22].

Another reason why it is all too easy to increase LogP in the early stage of drug discovery projects is that if the initial assay is a very specific target based assay (e.g., enzyme or artificially constructed complex) then as soon as hits have sufficient potency to be interesting, the screening cascade will require them to be looked at in cellular assays. In order to gain cellular potency it is all too easy to add lipophilicity as a quick way to get membrane permeability. Such compounds may have short-term benefits of demonstrating cellular activity but it is equally too easy to then just forge ahead with this "fattened ligand" in the desire to make further speedy progress

| Likelihood (50% probability) of permeability based onAZlogD lipophilicity for binned MW ranges ||
MW	AZlogD
<300	>0.5
300-350	>1.1
350-400	>1.7
400-450	>3.1
450-500	>3.4
>500	>4.5

Fig. 14.10 Lipophilicity required for maximising probability of passive permeability at different molecular weights (Redrawn for data in Ref. [11])

towards the project's next milestones. You should pause at this stage to re-evaluate the properties of the now current lead compounds. Of course, in previous eras of drug discovery when *in vivo* testing (or classical tissue pharmacology) was used much earlier in the discovery process this issue did not exist, because compounds that either showed no activity through lack of bioavailability or toxicity through inappropriate mode of action or side effects were dismissed or never found in the first place!

A study by Waring and colleagues of 9,598 AZ compounds has shown (Fig. 14.10) that on average larger molecules need more lipophilicity to be permeable through cell membranes [11]. Thus the apparent twin drivers of potency (MW = more interactions and LogP = increased permeability) are seen to be not truly independent variables in relation to bioavailability but increasingly linked as compounds get larger. In Lipinski's rule of five, the cutoff of MW of 500 is consistent with the experimentally observed upper limit of permeability of compounds through membranes without invoking active transport. What Waring's work now clearly shows is that the space below 500 is not binary, in the sense of being permeable with MW less than 500, but that it has an increasing LogP demand as 500 is approached.

14.4 Further Insight into Controlling These Addictions in Drug Discovery

While a typical medicinal chemistry publication will set out to show that a very logical process was followed towards achieving the project's objectives, it is clear that by any objective or subjective measure, the path that a drug discovery project actually follows through a multi-dimensional property space is non-linear. Figure 14.1 illustrated this by showing a pathway that could be taken from a start point X1 which has been found by some potency based screening process. From there on, there are numerous pathways (only one of which is illustrated) that could be taken to finally reach Xn, which is the candidate drug. Whether there is overlap of the zones in any given project as implied in the figure is only found out by trial and error, although experience with a related target may well inform on such target tractability.

Another important way to diminish the negative effects of potency is to be more realistic about what level of potency we should aspire to. Again surveys of literature data can help reset expectations. Analysis of a data set of known drugs by Overington et al. shows (Fig. 14.11) that the median affinity for current small-molecule drugs is ca. 20 nM ($pIC_{50} = 7.7$) (24) Unpublished in house data from GSK suggests that for oral drugs the affinity median is even less potent ($pIC_{50} = 6.7$). These average levels of potency for successful drugs are considerably lower than the often aspired to pK_i or pIC_{50} values of 9. Different target classes (e.g., ion channels vs. GPCRs) and whether agonists or antagonists (which will likely require differing levels of receptor occupancy for efficacy) will affect the aspirational potency at the outset of a project but as efficacy is, at the end of the day, what is needed, greater emphasis on the factors that can enable overall drug efficacy need to be more to the fore in the earliest stages rather than just potency at the target. Thus, while the desire for potency is understandable, the tendency to choose the most potent compounds in lead selection and then let it remain the primary driver through early stage lead optimization remains a strong and inappropriate attractor and must be resisted [24]. The fact that potency is often easy to measure (once the assay is established) can often mean that this is the data most likely first returned to the project team. The team (or individual) then tends to react to it by making synthesis decisions about what to make next without waiting for a fuller profile of data. So reducing the desire for potency in favor of better ADMET properties is

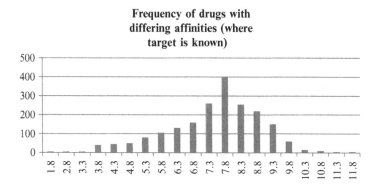

Fig. 14.11 Distribution of marketed drug affinities at target where target is known (Redrawn from data in Ref. [23])

one way of subjugating the potency attractor and its potentially fatal relationship to molecular obesity.

A useful tool to help maintain a balance of potency vs. ADMET properties is the Drug Efficiency concept, which tells you how much of your dose actually is available in the biophase of interest (DRUG$_{eff}$ = Biophase Concentration * 100/Dose) [25].

These authors more recently introduced the related Drug Efficiency Index DEI as a strategy towards low therapeutic dose (DEI = Log[DRUG$_{eff}$(%)] + pK$_D$) [26]. DEI is in effect a correction of the *in vitro* affinity (i.e. pK$_D$) by the *in vivo* pharmacokinetic potential. This simple descriptor directly connects efficacy and therapeutic dose with the potential to probe the balance between *in vitro* affinity and ADMET properties.

Finally a recent paper from Pfizer is worth highlighting [27]. It addresses the issue of what is the typical efficacious concentration (C$_{eff}$) of a drug that successfully passes through animal tox studies. For a series of 56 extensively studied compounds, the answer is <250 nM for total drug concentration and <40 nM for free drug. Interestingly 250 nM equates to 10 mg total dose in a human being, if we assume that we are just water! Clearly this is not the case, as we know drugs partition at an organ, tissue, cellular, organelle, lipid and target level. However it has long been known that idiosyncratic toxicology is rarely seen if the total daily dose of a drug is kept below 10 mg [28]. This only goes to emphasize the importance of getting the optimum balance between potency and bioavailability at the site of action. Traditionally bioavailability has been measured as the free concentration in plasma however this is not necessarily the free concentration inside a cell or some sub-cellular organelle. To enable this more appropriate measurement, MS based methods for understanding the local cellular concentration of drugs (as introduced by Per Artursson et al.) are evolving for use early in a drug discovery program and so there is increasingly no excuse for not being able to understand the issues on both sides of the balance before it is too late in a project [29].

Three excellent reviews that include further in depth discussion of physicochemical related attrition issues are recommended for further reading [30–33].

14.5 Summary

Molecular obesity and its inappropriate use to drive potency and get ligands through membranes have been killing too many drug discovery projects. Starting with the smallest possible lead (i.e. fragments) and striving to maintain their fitness through the use of various indices is now accepted as a key approach in a more holistic approach to contemporary drug discovery. The absolute need for potency should not be as dominant an attractor as we often allow it to become at the expense of other characteristics of a good drug.

There will always be some compounds that make it all the way to drugs and which lie outside of the known preferred space for likely success. However unless truly forced to by circumstances that are fully understood, it is not appropriate to set out with the mentality that my project will be the "exception that proves the rule" as a risk mitigation strategy!

References

1. Hann MM (2011) Molecular obesity, potency and other addictions in medicinal chemistry. Med Chem Commun 2:349–355
2. Kola I, Landis J (2004) Can the pharmaceutical industry reduce attrition rates? Nat Rev Drug Discov 3:711–716
3. Leeson PD, Springthorpe B (2007) The influence of drug-like concepts on decision-making in medicinal chemistry. Nat Rev Drug Discov 6:881–890
4. Lipinski CA, Lombardo F, Dominy BW, Feeney PJ (1997) Experimental and computational approaches to estimate solubility and permeability in drug discovery and development settings. Adv Drug Deliv Rev 46:3–26
5. Hughes JD et al (2008) Physiochemical drug properties associated with in vivo toxicological outcomes. Bioorg Med Chem Lett 18:4872–4875
6. Gleeson MP (2008) Generation of a set of simple, interpretable ADMET rules of thumb. J Med Chem 51:817–834
7. Ritchie TJ, Macdonald SJ (2009) The impact of aromatic ring count on compound developability – are too many aromatic rings a liability in drug design? Drug Discov Today 14:1011–1020
8. Young RJ, Green DVS, Luscombe CN, Hill AP (2011) Getting physical in drug discovery II: the impact of chromatographic hydrophobicity measurements and aromaticity. Drug Discov Today 16:822–830
9. Lovering F, Bikker J, Humblet C (2009) Escape from flatland: increasing saturation as an approach to improving clinical success. J Med Chem 52:6752–6756
10. Hann MM, Leach AR, Harper G (2001) Molecular complexity and its impact on the probability of finding leads for drug discovery. J Chem Inf Comput Sci 41(3):856–864
11. Waring MJ (2009) Defining optimum lipophilicity and molecular weight ranges for drug candidates – molecular weight dependent lower log D limits based on permeability. Bioorg Med Chem Lett 19:2844–2851
12. Hopkins AL, Groom CR, Alex A (2004) Ligand efficiency: a useful metric for lead selection. Drug Discov Today 9:430–431
13. Mortenson PN, Murray CW (2011) Assessing the lipophilicity of fragments and early hits. J Comput Aided Mol Des 25:663–667
14. Hopkins AL, Mason JS, Overington JP (2006) Can we rationally design promiscuous drugs? Curr Opin Struct Biol 16:127–137
15. Leach AR, Hann MM (2011) Molecular complexity and fragment-based drug discovery: ten years on. Curr Opin Chem Biol 15:489–496
16. Hann MM, Leach AR. Coping with complexity in molecular design. In: De novo molecular design. Wiley-VCH, Weinheim. ISBN 978-3-527-33461-2
17. Olsson TS, Williams MA, Pitt WR, Ladbury JE (2008) The thermodynamics of protein-ligand interaction and solvation: insights for ligand design. J Mol Biol 384:1002–1017
18. (A) Ferenczy GG, Keserű GM (2010) Thermodynamics guided lead discovery and optimization. Drug Discov Today 15:919. (B) Ferenczy GG, Keserű GM (2010) Enthalpic efficiency of ligand binding. J Chem Inf Model 50:1536
19. Hann MM, Keserű GM (2012) Finding the sweet spot: the role of nature and nurture in medicinal chemistry. Nat Rev Drug Discov 11:355–365

20. Selzer P, Roth HJ, Ertl P, Schuffenhauer A (2005) Complex molecules: do they add value? Curr Opin Chem Biol 9:310–316
21. Bajorath J (2002) Integration of virtual and high-throughput screening. Nat Rev Drug Discov 1:882–894
22. Nadin A, Hattotuwagama C, Churcher I (2012) Lead-oriented synthesis: a new opportunity for synthetic chemistry. Angew Chem Int Ed 51:1114–1122
23. Overington JP, Al-Lazikani B, Hopkins AL (2006) How many drug targets are there? Nat Rev Drug Discov 5:993–996
24. Perola E (2010) An analysis of the binding efficiencies of drugs and their leads in successful drug discovery programs. J Med Chem 53:2986–2997
25. Braggio S, Montanari D, Rossi T, Ratti E (2010) Drug efficiency: a new concept to guide lead optimization programs towards the selection of better clinical candidates. Expert Opin Drug Discov 5:609–618
26. Montanari D et al (2011) Application of drug efficiency index in drug discovery. Expert Opin Drug Discov 6:913–920
27. Wager TT et al (2013) Improving the odds of success in drug discovery: choosing the best compounds for in vivo toxicology studies. J Med Chem 56:9771–9779
28. Uetrecht J (2007) Idiosyncratic drug reactions: current understanding. Annu Rev Pharmacol Toxicol 47:513–539
29. Mateus A, Matsson P, Artursson P (2013) Rapid measurements of intracellular unbound drug concentrations. Mol Pharm 10:2467–2478
30. Waring MJ (2010) Lipophilicity in drug discovery. Expert Opin Drug Discov 5(3):235–248
31. Leeson PD, Empfield JR (2010) Reducing the risk of drug attrition associated with physicochemical properties. Annu Rep Med Chem 45:393–407
32. Gleeson MP, Hersey A, Montanari D, Overington J (2011) Probing the links between in vitro potency, ADMET and physicochemical parameters. Nat Rev Drug Discov 10:197–208
33. Hopkins AL et al (2014) The role of ligand efficiency metrics in drug discovery. Nat Rev Drug Discov 13:105–121

Chapter 15
Adventures in Small Molecule Fragment Screening by X-ray Crystallography

Joseph D. Bauman, Disha Patel, and Eddy Arnold

Abstract Since its conception in the early 1990s, fragment-based drug discovery (FBDD) has become established as a powerful tool for identifying new, chemically tractable pharmacophores. Unlike traditional methods that focus primarily on initial potency, FBDD stresses efficiency of binding and exploration of a highly diverse chemical space. Small fragment library sizes ($\sim$1,000 compounds) and the weak binding affinity of fragments have spurred the use of biophysical methods not readily applicable to screening of traditional compound libraries (greater than 100,000 compounds). X-ray crystallography is a powerful, yet under-appreciated, biophysical method for systematic identification of small molecule binding and discovery of potential inhibitory sites in a macromolecular target. Indeed, due to tremendous improvements in methodologies and technologies involved in X-ray data collection and analysis, it is now possible to collect data on a complete fragment library for a given macromolecular target during a single trip to a current generation synchrotron. Here we highlight some key insights and innovations learned from fragment screening campaigns targeting influenza and HIV-1 polymerases.

15.1 Introduction

Fragment-based drug discovery (FBDD) has become increasingly popular for detection of new pharmacophores for lead development. The small and less complex nature of fragments increases the probability of binding to a target protein, resulting in higher hit rates and an efficient search of chemical space [1–3]. Due to their small size, fragments often have weak potency relative to compounds screened in a traditional library. However, hits identified from fragment screening do not require deconstruction and can be efficiently optimized for specificity and potency.

Many biophysical and enzymatic screening assays have been developed for the detection of fragments that bind with low affinity, yet specifically, to a target

J.D. Bauman (✉) • D. Patel • E. Arnold (✉)
Center for Advanced Biotechnology and Medicine, Rutgers University, 679 Hoes Lane West, Piscataway, NJ 08854, USA
e-mail: bauman@cabm.rutgers.edu; arnold@cabm.rutgers.edu

© Springer Science+Business Media Dordrecht 2015   197
G. Scapin et al. (eds.), *Multifaceted Roles of Crystallography in Modern Drug Discovery*, NATO Science for Peace and Security Series A: Chemistry and Biology, DOI 10.1007/978-94-017-9719-1_15

protein. Historically, nuclear magnetic resonance (NMR) spectroscopy and X-ray crystallography were used for primary screening. More recently, surface plasmon resonance and differential scanning fluorimetry (thermal shift) have also emerged as rapid and powerful methods for hit identification. However, crystallography remains unmatched in its ability to ascertain the three-dimensional structures of the ligand(s) bound to the target protein for structure-based drug design (SBDD). When crystals of the target protein of sufficient quantity and quality are obtainable, X-ray crystallographic fragment screening (XCFS) can be an excellent tool for drug discovery as well as for biochemical characterization of the protein of interest. For crystals to be appropriate for XCFS they should meet the following criteria (adapted from reference [4]):

1. Significant quantities of crystals must be obtainable that can routinely diffract X-rays to high resolution. A cutoff of 2.5 Å can be considered reasonable as crystal fragment soaking often decreases the diffraction resolution of the crystals. However, <2.0 Å is strongly preferred to ensure unambiguous fitting of fragments into electron density.
2. The protein must be in a biologically relevant conformation.
3. The druggable sites must not be occluded by protein-protein crystal contacts, a natural ligand, or a chemical used for crystallization or cryoprotection.
4. The crystals must be robust enough to survive the soaking of fragments.
5. The pH and ionic strength of the crystallization solution should be near physiological conditions to ensure biological relevant interactions.

In addition, crystals with higher symmetry and a small unit cell are preferable to minimize data collection and analysis time [5]. Despite this, X-ray crystallography is often under-appreciated as a primary screening method as it is associated with being relatively low throughput, highly resource intensive, and very specialized.

Recent and continuing advancements in crystal storage/transport, robotic crystal handling, intense X-ray sources, automated data collection software, and sensitive/fast detectors have allowed for data collection at rates unimagined at the beginning of the fragment screening field. Technical improvements and optimized strategies for data collection and processing allows for screening of 50–80 compounds per hour as cocktails consisting of five compounds per crystal. It is, therefore, possible to screen a sizable fragment library of a 1,000 compounds in a single visit to a high intensity X-ray facility. X-ray crystallography remains attractive as a primary screen since it not only identifies fragment binding to the target protein, but also knowledge of the detailed three-dimensional structure facilitates rapid SBDD.

15.2 Library Design

As with any screening methodology, the success of the approach is dependent on the composition of the library to be tested. Computational filters can be used to either design a general screening library or further restricted for a more focused library targeting a specific class of protein [6–8]. Astex Pharmaceuticals, a pioneer in FBDD, proposed the Rule of Three (Ro3) to guide library design, based on a molecular mass <300 Da, $\leq$ three hydrogen bond donors, $\leq$ three hydrogen bond acceptors, calculated logP (clogP) <3, rotatable bonds $\leq$3, and a polar surface area $\leq$60 Å^2 [9, 10]. Ro3 has recently attracted criticism within the field for being too limiting. During a screening campaign against endothiapepsin using a fragment library not designed using the Ro3 guidelines, Koster et al. found only four out of the eleven hits identified from the screening actually satisfied the Ro3 criteria while the remaining seven fragments were found to violate at least one of the Ro3 guidelines [11]. The backlash against Ro3 has led to a nickname of "the Voldemort rule" by some leaders in the FBDD field [12] and should not be considered as absolute as recently pointed out by some of the authors of the original manuscript [13].

15.3 Screening Methodologies

For certain biophysical approaches, including X-ray crystallography, it is necessary to screen cocktails or mixtures of fragments to improve throughput. In XCFS, fragment cocktails are soaked into preformed crystals for minutes to several hours at concentrations of 10 mM or higher. The high concentration of fragments is necessary to ensure maximum occupancy for low affinity ligands. Early cocktail design relied on large numbers of fragments per cocktail (16–150 compounds), but this approach was restricted by the solubility of individual fragments, difficulties in identifying the bound fragment, and perhaps most critically, challenges with predicting reactions between fragments as well as the constituents of the crystallization condition [14, 15]. Due to these issues, it is now common for cocktail size to be limited to four or five compounds [16, 17].

Hit identification is based on optimizing the fit of the electron density to the individual fragments within the cocktail. To facilitate deconvolution, cocktails are often designed to maximize shape and chemical diversity [14, 16]. An alternative approach is to minimize diversity in a cocktail and thereby minimize the likelihood of chemical reactions between the fragments [17]. When diversity is minimized, deconvolution relies on high-resolution and well-ordered electron density to determine the fragment identity. Unfortunately, it is often the case that the electron density for the initial fragment hit is poor and derivatization of the fragment is necessary to improve the electron density. The optimal cocktail design continues to evolve as new methodologies are developed.

Co-crystallization of the fragment with the protein can be used successfully for screening [18]. However, this approach to XCFS is very labor and resource intensive. Fragment precipitation and possible chemical reactivity with the crystallization condition can limit the applicability of co-crystallization as a viable approach. Furthermore, it may be necessary to explore alternate crystallization conditions since fragment binding can alter the conformational dynamics of the protein. Co-crystallization is only used when fragment binding damages pre-formed crystals or when crystal soaking fails to identify hits [19].

15.4   Experiences from Fragment Screening by X-ray Crystallography

The Arnold laboratory has successfully conducted three screening campaigns two of which—the HIV-1 reverse transcriptase (RT) and influenza endonuclease (PA$_N$)—have been recently published [20, 21]. A sparse fragment library was assembled for the screening campaigns as follows: 500 compounds purchased from Marbridge (Cornwall, UK), 175 compounds based on the published recommendations of Christophe Verlinde and Wim Hol [14], and an additional 100 compounds were generously gifted by James Williamson (The Scripps Research Institute, La Jolla, unpublished). To expedite the screening process, the fragments were grouped into 143 cocktails of four to eight compounds (100 mM each in d$_6$-DMSO). As fragment hits were discovered, derivatives were purchased based on structural similarities to the initial hit and *in silico* docking results. In addition, to identifying promising fragments for lead development, the screening campaigns also revealed several valuable insights, some of which are highlighted in the following sections.

15.4.1   Lessons Learned from Fragment Screening against HIV-1 RT

HIV-1 reverse transcriptase (RT) has been well characterized structurally and biochemically due to its importance in HIV viral replication and, in turn, as one of medicine's central drug targets. RT is a heterodimer of p66 and p51 subunits with the p66 subunit containing the fingers, palm, and thumb polymerase subdomains and the C-terminal RNase H domain linked by a connection subdomain to the polymerase (Fig. 15.1). RT remains a major drug target in the fight against AIDS, with 13 of the 26 approved anti-AIDS drugs inhibiting RT. Unfortunately, the continued development of resistance to the currently approved drugs necessitates the development of new drugs that target new sites to avoid cross-resistance.

15.4.1.1   Crystal Optimization

Crystal diffraction quality proved to be a major challenge for initiating a fragment-screening campaign against HIV-1 reverse transcriptase (RT). Crystal engineering was undertaken to improve the resolution quality of RT while ensuring the protein remained in a biologically relevant conformation [22]. Crystals of the apo RT69A construct identified from the engineering efforts routinely diffracted X-rays to 1.85 Å. However, these crystals were not robust for fragment soaking, with only one out of ten crystals surviving soaks for a particular cocktail and producing a reasonable quality dataset. The poor crystal stability was speculated to be the result

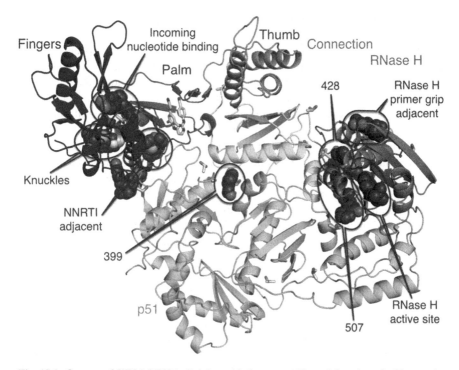

Fig. 15.1 Cartoon of HIV-1 RT52A-rilpivine with fragments. The subdomains of p66 are color coded as fingers (*blue*), palm (*red*), thumb (*green*), connection (*yellow*), and RNase H (*orange*). Rilpivirine (*wheat* and *blue sticks*), DMSO (*wheat*, *red*, and *yellow sticks*), bound fragments (*purple spheres*) are also depicted (Color figure online)

of binding to an allosteric binding site in HIV-1 RT, referred to as the non-nucleoside reverse transcriptase inhibitor (NNRTI) binding pocket. As a result, an alternate construct, RT52A, co-crystallized with a potent NNRTI drug, rilpivirine, was used for screening. Crystals of the RT52A-rilpivirine complex routinely diffracted X-rays to 1.80 Å and were highly robust during fragment soaking with 93 % of crystals surviving the fragment soaks.

15.4.1.2 Optimization of Soaking Condition

When possible, making a single solution for both fragment soaking and cryoprotection is preferred to minimize labor and resources during screening. Using a single solution for soaking and cryoprotection also minimizes crystal damage due to physical handling. The DMSO that the fragments are stored in can also act as a potent cryoprotectant and, in the case, of RT-rilpivirine crystals, 20 % (v/v) DMSO was used with 7 % (v/v) ethylene glycol as sufficient cryoprotectant.

15.4.1.3   Improving Fragment Solubility with L-Arginine

At the high concentrations used for XCFS, fragment solubility, especially in the crystallization condition, is a significant concern. With most other biophysical techniques, compound aggregation and precipitation can invalidate the results due to non-specific interactions and a loss of sufficiently soluble compound for binding. In crystallographic screening, it is not required for the fragment to be fully soluble at the tested concentration, as it is possible to recover crystals from drops that are only partially transparent due to precipitation. The effective fragment concentration is still important since fragment binding is normally quite weak due to the smaller number of atoms interacting with the protein. In order to improve fragment solubility and in turn its effective concentration, L-arginine at a concentration of 80 mM was found to be an valuable additive during soaking [20]. We hypothesize that the guanidinium group of L-arginine improves fragment solubility by forming π-stacking as well as hydrogen-bonding interactions with fragments [23].

15.4.1.4   Cryoprotection with Trimethylamine-N-Oxide

During the fragment-screening campaign targeting RT, trimethylamine-N-oxide (TMAO) was found to improve the diffraction quality of crystals. Soaking at a concentration of 6 % (w/v) TMAO improved the resolution of the HIV-1 RT crystals from 1.80 to 1.51 Å [24]. A similar improvement in diffraction resolution with TMAO was also observed for crystals of influenza endonuclease and other HIV-1 RT complexes (unpublished). Based on these results, an improvement in X-ray diffraction quality was also observed for crystals of alphavirus polyprotein using 5 % (w/v) TMAO during cryoprotection [25]. Independently of our results, Mueller-Dieckmann et al. found that 4 M TMAO in water by itself can act as an excellent cryoprotective agent [26]. Improvement in HIV-1 RT crystal diffraction resolution was observed solely through soaking of TMAO, not through co-crystallization. Importantly, although useful in improving diffraction quality of some crystals, TMAO was found to decrease fragment density in the crystals. Visual precipitation of the fragment soaking solution in the presence of TMAO during soaking may indicate a decrease in fragment solubility. The resulting decrease in effective fragment concentration, in turn, may be responsible for the loss in binding.

15.4.1.5   Importance of a Blank Dataset

The presence of high DMSO concentrations during cryoprotection necessitates use of a reference dataset, consisting of a blank dataset measured from a crystal without the presence of any fragments. Once the bound DMSO molecules have been located, the new structure and dataset can be used as a reference for detection of fragment binding. For RT52A-rilpivirine, the 20 % (v/v) DMSO dataset revealed 16 DMSO binding sites throughout the protein and proved to be valuable for distinguishing fragment from solvent binding (Fig. 15.1). $F_{obs(fragment\ soaked)}$-$F_{obs(DMSO\ blank)}$ (F_o-F_o) difference maps were used to identify fragment binding directly and/or through side-chain and backbone movements. The location of bound DMSO molecules is also informative if binding occurs at a known druggable site, particularly one of great interest. An alternative solvent may be used, such as methanol, to avoid competition of the very high concentration solvent with fragments.

15.4.1.6 High-Throughput Data Collection and Processing

Unlike typical X-ray data collection strategies, fragment screening stresses rapid data collection above collection of the best possible dataset for every crystal. Depending on the crystals, it may be possible to collect a dataset in half the time with only a moderate decrease in resolution. For the RT52A-rilpivirine crystals, decreasing the collection time in half resulted in a decrease in resolution for a typical crystal from 1.85 to 2.0 Å resolution. By rapidly processing the data and computing the F_o-F_o difference maps, it is possible to collect a higher resolution dataset for crystals when fragment binding is detected. The need for a high-speed data collection pass has now been diminished by the presence of very rapid pixel array detectors at certain beamlines.

15.4.1.7 Halogens

Halogenated fragments are often under-represented in fragment libraries due to molecular mass cutoffs. Halogens have been used for rapid hit identification during X-ray crystallographic and NMR screening. Specifically, anomalous signal from bromine atoms [27] and ^{19}F chemical shift for NMR spectroscopy [28] can be used for detection of binding. Bromine can also be used as a convenient chemical handle for elaboration of the initial hit.

An unexpected preference for halogenated fragments was discovered during the RT fragment screening campaign. Compared to the overall 4.4 % hit rate for fragments in the library, 23.5 % (4 out of 17 fragments) were brominated fragments while the fluorinated fragment hit rate was 24.1 % (7 out of 29 fragments). The higher hit rate for halogen-containing fragments indicated that a halogenated-focused fragment library may be beneficial. Based on these results, Tiefenbrunn et al. screened a library of 68 brominated fragments against HIV-1 protease with a hit rate of four to eight times higher than when a non-halogenated library was screened [29].

15.4.1.8 Discovery of New Allosteric Binding Sites on HIV-1 RT

Seven new sites were discovered throughout RT by fragment screening. In the polymerase region, we discovered the Incoming Nucleotide Binding, Knuckles,

NNRTI adjacent, and 399 sites while the 428, RNase H Primer Grip Adjacent, and 399 sites reside in the RNase H region (Fig. 15.1). Subsequent enzymatic assays indicated that fragments bound at the Incoming Nucleotide Binding, Knuckles, and NNRTI adjacent sites were inhibitory, indicating potential for further drug development.

Crystallographic fragment screening allowed for chemical probing of the conformational dynamics of RT. Although the crystal lattice constrains the protein from large conformational changes, it allows for sufficient flexibility to detect modest movements. Fragment binding to RT allowed for stabilization of a previously undetected pocket referred to as the Knuckles site located between the fingers and palm subdomains. In the Knuckles open conformation there is a 1.2 Å backbone movement in the fingers subdomain and a $\sim$3.0 Å deformation in the nearby incoming nucleotide-binding site. Apparently by changing RT's conformational dynamics, a fragment bound at the site inhibited the enzyme at 600 μM [20].

15.4.2 Fragment Screening of Influenza Endonuclease

Influenza transcription is primed using the 5' 11–13 nucleotides, which includes the 5' mRNA cap, of cellular mRNA acquired through a process referred to as "cap snatching". The cleavage of the cellular mRNA is performed by the influenza cap-snatching endonuclease, which resides in the N-terminal domain of the polymerase PA (PA_N) subunit [30, 31]. The original crystal forms of PA_N were not appropriate for XCFS due to difficulty in reproducing the crystals [32] or having an occluded active site [31]. To overcome these limitations, crystal engineering was applied to make readily reproducible crystals that diffracted X-rays to high resolution and have an open active site [21].

15.4.2.1 Detection of a Third Metal at the Active Site

Previous studies through biochemistry and crystallography indicated one or two metals at the active site of PA_N [30, 31, 33, 34] but XCFS detected a third metal coordinated in two different fragment soaks. Based on these results, various metals were titrated into crystals to determine if metal binding could occur without coordination by a fragment. A 100 mM solution of Ca^{2+} revealed binding at the third metal site indicating that metal binding at this site is not simply an artifact of fragment binding [21]. Three-metal ion coordination has been observed in other endonucleases and has been proposed to stabilize the negative charge of the transition state of the nucleic acid and recruit a water molecule to protonate the leaving group [35–40]. Further studies are required to determine if this is the case for the influenza endonuclease (Fig. 15.2).

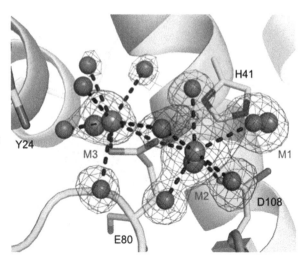

Fig. 15.2 Electron density at the PA$_N$ active site after a 100 mM CaCl$_2$ soak. Previously detected metals are shown as *green spheres* while modeled calcium ions are *yellow spheres*. Electron density is shown for a omit map (2.5 σ contour) with the metals and coordinating water (*red spheres*) removed (Color figure online)

15.4.2.2 Fragments and Multiple Binding

Other biophysical techniques often discard multiple binders as having non-specific interactions with the protein. Since X-ray crystallography allows visualization of the specific interactions between a fragment and multiple binding sites within a protein, it is possible to improve the compound specificity through chemical modifications. This was demonstrated for a fragment hit that bound to three sites within PA$_N$ but was quickly optimized to a single binding mode at the active site (Fig. 15.3). A combination of X-ray crystallography, enzymatic assays, molecular docking, and medicinal chemistry quickly developed the hydroxypyridinone series from a 16 μM initial hit to an compound with 11 nM potency and significant antiviral activity in cellular assays [21, 41].

15.5 Conclusions

When suitable crystals of the target protein are obtainable, fragment screening by X-ray crystallography has been established to be a valuable tool for drug discovery and development. In addition, it can also provide exceptional insights into functions and characteristics of the target protein. XCFS can provide previously unseen snapshots of protein conformation as well as probe function and mechanism. The continued development of high-throughput crystallography and its unique benefits for structure-based drug design make crystallographic fragment screening an excellent tool for characterization of target proteins.

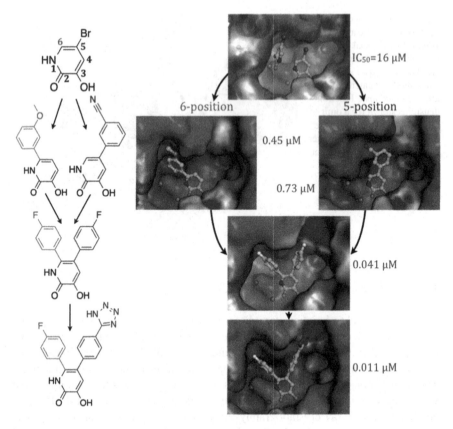

Fig. 15.3 Structure-based lead optimization of 5-chloro-2,3-dihydroxypyridine. The initial fragment hit was detected bound at three positions at or near the endonuclease active site. The electrostatic surface of the flexible active site cleft is shown (APBS) for each structure. Rapid determination of crystal structures of the compounds bound to the endonuclease allowed for a greater than a thousand-fold increase in potency with less than 100 compounds synthesized

Acknowledgement EA is grateful to the National Institutes of Health for support from grants R37 AI027690 (MERIT Award) and P50 GM103368. We also thank our collaborators in RT studies, both past and present.

References

1. Hajduk PJ, Greer J (2007) A decade of fragment-based drug design: strategic advances and lessons learned. Nat Rev Drug Discov 6:211–219
2. Hesterkamp T, Whittaker M (2008) Fragment-based activity space: smaller is better. Curr Opin Chem Biol 12:260–268
3. Erlanson DA, Wells JA, Braisted AC (2004) Tethering: fragment-based drug discovery. Annu Rev Biophys Biomol Struct 33:199–223

4. Bauman JD, Patel D, Arnold E (2012) Fragment screening and HIV therapeutics. Top Curr Chem 317:181–200
5. Begley DW, Davies DR, Hartley RC, Hewitt SN, Rychel AL, Myler PJ, Van Voorhis WC, Staker BL, Stewart LJ (2011) Probing conformational states of glutaryl-CoA dehydrogenase by fragment screening. Acta Crystallogr Sect F: Struct Biol Cryst Commun 67:1060–1069
6. Howard N, Abell C, Blakemore W, Chessari G, Congreve M, Howard S, Jhoti H, Murray CW, Seavers LCA, van Montfort RLM (2006) Application of fragment screening and fragment linking to the discovery of novel thrombin inhibitors. J Med Chem 49:1346–1355
7. Bodoor K, Boyapati V, Gopu V, Boisdore M, Allam K, Miller J, Treleaven WD, Weldeghiorghis T, Aboul-ela F (2009) Design and implementation of an ribonucleic acid (RNA) directed fragment library. J Med Chem 52:3753–3761
8. Wang Y-S, Strickland C, Voigt JH, Kennedy ME, Beyer BM, Senior MM, Smith EM, Nechuta TL, Madison VS, Czarniecki M, McKittrick BA, Stamford AW, Parker EM, Hunter JC, Greenlee WJ, Wyss DF (2010) Application of fragment-based NMR screening, X-ray crystallography, structure-based design, and focused chemical library design to identify novel microM leads for the development of nM BACE-1 (beta-site APP cleaving enzyme 1) inhibitors. J Med Chem 53:942–950
9. Congreve M, Carr R, Murray C, Jhoti H (2003) A "rule of three" for fragment-based lead discovery? Drug Discov Today 8:876–877
10. Congreve M, Chessari G, Tisi D, Woodhead AJ (2008) Recent developments in fragment-based drug discovery. J Med Chem 51:3661–3680
11. Köster H, Craan T, Brass S, Herhaus C, Zentgraf M, Neumann L, Heine A, Klebe G (2011) A small nonrule of 3 compatible fragment library provides high hit rate of endothiapepsin crystal structures with various fragment chemotypes. J Med Chem 54:7784–7796
12. Zartler T (2013) What's the fire behind the smoke? http://practicalfragments.blogspot.com/2013/04/whats-fire-behind-smoke.html. Accessed 9 Apr 2014
13. Jhoti H, Williams G, Rees DC, Murray CW (2013) The "rule of three" for fragment-based drug discovery: where are we now? Nat Rev Drug Discov 12:644–645
14. Verlinde CLMJ, Fan E, Shibata S, Zhang Z, Sun Z, Deng W, Ross J, Kim J, Xiao L, Arakaki T, Bosch J, Caruthers JM, Larson ET, LeTrong I, Napuli A, Kelley A, Mueller N, Zucker F, Van Voorhis WC, Buckner FS, Merritt EA, Hol WGJ (2009) Fragment-based cocktail crystallography by the Medical Structural Genomics of Pathogenic Protozoa Consortium. Curr Top Med Chem 9:1678–1687
15. Nienaber VL, Richardson PL, Klighofer V, Bouska JJ, Giranda VL, Greer J (2000) Discovering novel ligands for macromolecules using X-ray crystallographic screening. Nat Biotechnol 18:1105–1108
16. Hartshorn M, Murray C, Cleasby A, Frederickson M, Tickle I, Jhoti H (2005) Fragment-based lead discovery using X-ray crystallography. J Med Chem 48:403–413
17. Spurlino JC (2011) Fragment screening purely with protein crystallography. Methods Enzymol 493:321–356
18. Perryman AL, Zhang Q, Soutter HH, Rosenfeld R, McRee DE, Olson AJ, Elder JE, David Stout C (2010) Fragment-based screen against HIV protease. Chem Biol Drug Des 75:257–268
19. Davies TG, Tickle IJ (2011) Fragment screening using X-Ray crystallography. Top Curr Chem 317:33–59
20. Bauman JD, Patel D, Dharia C, Fromer M, Ahmed S, Frenkel Y, Eck JT, Ho W, Das K, Shatkin A, Arnold E (2013) Detecting allosteric sites of HIV-1 reverse transcriptase by X-ray crystallographic fragment screening. J Med Chem 56:2738–2746
21. Bauman JD, Patel D, Baker SF, Vijayan RSK, Xiang A, Parhi AK, Martínez-Sobrido L, Lavoie EJ, Das K, Arnold E (2013) Crystallographic fragment screening and structure-based optimization yields a new class of influenza endonuclease inhibitors. ACS Chem Biol 8:2501–2508
22. Bauman JD, Das K, Ho WC, Baweja M, Himmel DM, Arthur D, Clark J, Oren DA, Boyer PL, Hughes SH, Shatkin AJ, Arnold E (2008) Crystal engineering of HIV-1 reverse transcriptase for structure-based drug design. Nucleic Acids Res 36:5083–5092

23. Flocco MM, Mowbray SL (1994) Planar stacking interactions of arginine and aromatic side-chains in proteins. J Mol Biol 235:709–717
24. Kuroda DG, Bauman JD, Challa JR, Patel D, Troxler T, Das K, Arnold E, Hochstrasser RM (2013) Snapshot of the equilibrium dynamics of a drug bound to HIV-1 reverse transcriptase. Nat Chem 5:174–181
25. Shin G, Yost SA, Miller MT, Elrod EJ, Grakoui A, Marcotrigiano J (2012) Structural and functional insights into alphavirus polyprotein processing and pathogenesis. Proc Natl Acad Sci U S A 109:16534–16539
26. Mueller-Dieckmann C, Kauffmann B, Weiss MS (2011) Trimethylamine N-oxide as a versatile cryoprotective agent in macromolecular crystallography. J Appl Crystallogr 44:433–436
27. Blaney J, Nienaber V, Burley SK (2006) Fragment-based lead discovery and optimisation using X-ray crystallography, computational chemistry, and high-throughput organic synthesis. In: Jahnke W, Erlanson DA (eds) Fragment-based approaches in drug discovery. Methods and principles in medicinal chemistry. Wiley–VCH, Weinheim, pp 215–248
28. Dalvit C, Flocco M, Veronesi M, Stockman BJ (2002) Fluorine-NMR competition binding experiments for high-throughput screening of large compound mixtures. Comb Chem High Throughput Screen 5:605–611
29. Tiefenbrunn T, Forli S, Happer M, Gonzalez A, Tsai Y, Soltis M, Elder JH, Olson AJ, Stout CD (2014) Crystallographic fragment – based drug discovery: use of a brominated fragment library targeting HIV protease. Chem Biol Drug Des 83:141–148
30. Yuan P, Bartlam M, Lou Z, Chen S, Zhou J, He X, Lv Z, Ge R, Li X, Deng T, Fodor E, Rao Z, Liu Y (2009) Crystal structure of an avian influenza polymerase PA(N) reveals an endonuclease active site. Nature 458:909–913
31. Dias A, Bouvier D, Crépin T, McCarthy AA, Hart DJ, Baudin F, Cusack S, Ruigrok RWH (2009) The cap-snatching endonuclease of influenza virus polymerase resides in the PA subunit. Nature 458:914–918
32. Dubois RM, Slavish PJ, Baughman BM, Yun M-K, Bao J, Webby RJ, Webb TR, White SW (2012) Structural and biochemical basis for development of influenza virus inhibitors targeting the PA endonuclease. PLoS Pathog 8:e1002830
33. Doan L, Handa B, Roberts NA, Klumpp K (1999) Metal ion catalysis of RNA cleavage by the influenza virus endonuclease. Biochemistry 38:5612–5619
34. Crépin T, Dias A, Palencia A, Swale C, Cusack S, Ruigrok RWH (2010) Mutational and metal binding analysis of the endonuclease domain of the influenza virus polymerase PA subunit. J Virol 84:9096–9104
35. Syson KK, Tomlinson CC, Chapados BRB, Sayers JRJ, Tainer JAJ, Williams NHN, Grasby JAJ (2008) Three metal ions participate in the reaction catalyzed by T5 flap endonuclease. J Biol Chem 283:28741–28746
36. Ivanov I, Tainer JA, McCammon JA (2007) Unraveling the three-metal-ion catalytic mechanism of the DNA repair enzyme endonuclease IV. Proc Natl Acad Sci 104:1465–1470
37. Sam MDM, Perona JJJ (1999) Catalytic roles of divalent metal ions in phosphoryl transfer by EcoRV endonuclease. Biochemistry 38:6576–6586
38. Kovall RA, Matthews BW (1999) Type II restriction endonucleases: structural, functional and evolutionary relationships. Curr Opin Chem Biol 3:578–583
39. Horton NC, Newberry KJ, Perona JJ (1998) Metal ion-mediated substrate-assisted catalysis in type II restriction endonucleases. Proc Natl Acad Sci U S A 95:13489–13494
40. Horton NC, Perona JJ (2004) DNA cleavage by EcoRV endonuclease: two metal ions in three metal ion binding sites. Biochemistry 43:6841–6857
41. Parhi AK, Xiang A, Bauman JD, Patel D, Vijayan RSK, Das K, Arnold E, Lavoie EJ (2013) Phenyl substituted 3-hydroxypyridin-2(1H)-ones: inhibitors of influenza A endonuclease. Bioorg Med Chem 21:6435–6446

Chapter 16
Structure-Based Drug Design to Perturb Function of a tRNA-Modifying Enzyme by Active Site and Protein-Protein Interface Inhibition

Gerhard Klebe

Abstract Drug research increasingly focuses on the interference with protein-protein interface formation as attractive opportunity for therapeutic intervention. The tRNA-modifying enzyme Tgt, a putative drug target to fight Shigellosis, is only functionally active as a homodimer. To better understand the driving forces responsible for assembly and stability of the formed homodimer interface we embarked onto a computational and mutational analysis of the interface-forming residues. We also launched spiking ligands into the interface region to perturb contact formation. We controlled by non-degrading mass spectrometry the actual ratio of monomer-dimer equilibrium in solution and used crystal structure analysis to elucidate the geometrical changes resulting from the induced perturbance. A patch of four aromatic amino acids, embedded into a ring of hydrophobic residues and further stabilized by a network of H-bonds is essential for the dimer contact. Apart from the aromatic hot spot, the interface shows an extended loop-helix motif, which exhibits remarkable flexibility. In the destabilized mutant variants and the complexes with the spiking ligands, the loop-helix motif adopts deviating conformations in the interface region. This motivated us to follow a strategy to raise small molecule binders against this motif to mould the loop geometry in a conformation incompatible with the interface formation.

16.1 Introduction

Shigella dysentery is a severe diarrheal illness, caused by *Shigella* bacteria. These bacteria are ingested with contaminated water or food and adhere to epithelial cells in the intestinal mucosa. They are extremely contagious, 10–100 bacteria are enough

G. Klebe (✉)
Department of Pharmaceutical Chemistry, University of Marburg, Marbacher Weg 6,
D35032 Marburg, Germany
e-mail: klebe@mailer.uni-Marburg.de

© Springer Science+Business Media Dordrecht 2015
G. Scapin et al. (eds.), *Multifaceted Roles of Crystallography in Modern Drug Discovery*, NATO Science for Peace and Security Series A: Chemistry and Biology,
DOI 10.1007/978-94-017-9719-1_16

to cause an infection. Worldwide, shigellosis represents a serious problem. Almost 170 million cases are reported annually, from which over a million are fatal. The disease is widespread in developing countries, but over 1.5 million cases are also annually reported in industrialized countries. Above all, the disease flourishes under conditions of inadequate hygiene and poor water quality as is found in war, natural catastrophes, famine, and in refugee camps. Dysentery is a particular problem in Africa where it can occur concomitantly with AIDS [1].

As with any bacterial infectious disease, shigellosis can be treated with antibiotics. The infections that occur in industrialized countries are cured in this way. Unfortunately *Shigella*, which are very similar to *Escherichia coli* that naturally occur in the intestinal flora, have a tendency to become very quickly resistant to antibiotics. Moreover, an antibiotic therapy also kills the naturally occurring bacteria of the intestinal flora, and this also produces diarrheal symptoms and severe dehydration in the patients. This can lead to a life-threatening disruption of the electrolyte homeostasis, particularly in young children. Therefore, specific therapeutic approaches are sought that suppress the pathogenesis of the shigellosis.

16.2 Involvement of the tRNA-Modifying Enzyme TGT in the Infection Mechanism

Shigella attack the epithelial cells in the intestines. To gain entrance to these cells, the bacteria produce their own virulence factors, so-called invasins. These are proteins that form a sophisticated apparatus with the proteins on the epithelial cells, which allow the penetration and proliferation of the bacteria in the infected cells. The genes for the virulence factors are coded on a plasmid. Their expression in cases of infection is regulated by different transcription factors, particularly *VirF* is responsible for the pathogenesis of the bacteria, and altered tRNA molecules are needed so that it can be efficiently synthesized in the ribosome. tRNA is loaded with an amino acid at its terminus that corresponds to the base-pair triplet in the anticodon loop. The genetic information encoded in the base-triplet of the mRNA is transferred when the mRNA binds to the corresponding tRNA in the ribosome during translation. This tRNA carries the right amino acid so that the growing peptide chain of the nascent protein is correctly constructed. The changes in the required tRNA affect the base in position 34 of the so-called wobble region where a modified base must be incorporated. If these changes do not occur, the translation is inefficient. *Shigella* could then barely produce enough of the needed invasins to infect the epithelial cells. Their pathogenic potential is therefore severely reduced.

The bacteria have enzymes that can carry out these changes in the tRNA. In the first step, a guanine is cut out of the tRNA molecule at position 34 and replaced with an altered base, preQ$_1$ (Fig. 16.1). This step is catalyzed by tRNA-guanine transglycolase (TGT). The exchanged base in the tRNA is further modified in the next step of the enzymatic cascade so that the base queuine is obtained as

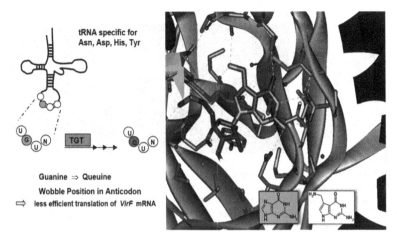

Fig. 16.1 The modified base queuine (Q) is incorporated into the wobble position of tRNA replacing guanine (G). The reaction occurs in bacteria stepwise and starts with the incorporation of preQ$_1$. On the right, the catalytic center of the enzyme is shown with part of the substrate tRNA and preQ$_1$. Inhibition of TGT, a glycosylase, makes translation of *VirF*, coding for invasion proteins, inefficient and thus reduces pathogeniety of *Shigella* bacteria

final product. TGT inhibitors therefore represent a specific therapeutic principle to selectively attack the development of pathogeniety of *Shigella*. In contrast to a therapy with broad-spectrum antibiotics, the bacteria are not killed but rather the disease-causing infection of the epithelial cells is prevented. Higher-developed eukaryotic organisms also have such an enzyme. In contrast to the bacteria, which use a homodimer, the eukaryotic enzyme is a heterodimer. Moreover the higher-developed organisms do not transform preQ$_1$ to the end-product queuine but incorporate the latter queuine directly into the tRNA.

16.3 tRNA Modifications Catalyzed by TGT

First, the crystal structure of TGT in complex with preQ$_1$ had to be determined in a related species [2, 3]. This shows an exchange of a Phe for a Tyr in the active site, which is immaterial for the binding. The structure in complex with a part of the tRNA was elucidated (Fig. 16.2).

Initially, the tRNA binds with the covalently attached guanine. The base with its ribose moiety is pulled out of the tRNA molecule and is specifically recognized by Asp102, Asp156, Gln203, Gly230, and Leu231. The reaction starts with a nucleophilic attack at carbon C1 of the ribose ring. The C1-N bond is broken and guanine is released. The base leaves the binding pocket with a water molecule, and preQ$_1$ is taken up into the same binding site. For this, the peptide bond between

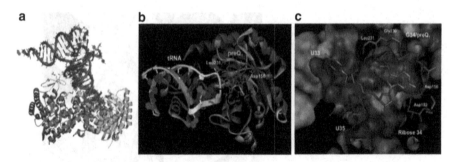

Fig. 16.2 (**a**) Structure of the TGT homodimer with bound tRNA. One subunit (*right, light gray*) contributes the catalytic center whereas the second monomer (*left, dark gray*) is only used to orient the tRNA properly for the enzymatic reaction. (**b**) One monomer with bound tRNA and the catalytic center. (**c**) Catalytic center with four subpockets. The exchanged base preQ$_1$ is recognized in the guanine pocket (G34/preQ1). In the proceeding open bowl-shaped pocket (U33) U33 of the substrate is recognized, in the pocket (U35) the subsequent U35 is accommodated. The ribose pocket (Ribose 34) hosts the sugar moiety of the central nucleotide

Leu231 and Ala232 must flip over. The basic nitrogen atom of preQ$_1$ then releases a proton and carries out a nucleophilic attack on the ribose, which is covalently bound to Asp280. Once the new bond to the tRNA is formed, the altered tRNA leaves the enzyme. Asp102 is critically involved in the recognition process of the bound base [4].

16.4   Development of Active Site Inhibitors

Active site inhibitor design started in the guanine pocket using de novo design, docking and virtual screening [3, 5]. Several small heterocyclic compounds were discovered. Among these the *lin*-benzoguanine moiety appeared most promising (Fig. 16.3) [6]. Substitution at 2-position allows addressing the U33 subpocket. Compounds with single-digit nanomolar inhibition could be produced [7]. In the solvent-exposed pocket many of the developed inhibitors adopt binding modes showing the attached substituent scattered over several orientations. In the crystal structures pronounced disorder is detected. Substituents attached in 4-position accommodate the ribose subpocket and the corresponding ligands also experience nanomolar binding [8]. The substituents have to cross a cluster of several water molecules and favorable binding is only successful if the substituents place a polar atom in the center of the perturbed water cluster [9]. Combining the 2- and 4-substituents results in ligands with subnanomolar potency [10]. Interestingly enough, the 4-substituents place their terminal atoms in the ribose subpocket close the dimer interface. This stimulated us to design and develop inhibitors with long spiking substituents launching into the interface region (s. below).

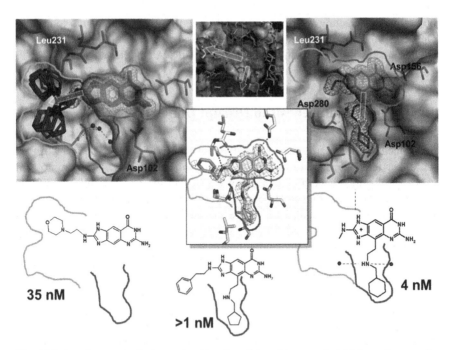

Fig. 16.3 Attachment of substituents in either 2- and 4-position reveals inhibitors of nanomolar potency. They address either the U33 (*left*) or ribose (*right*) subpocket. Combining both substituents in one molecule results in sub-nanomolar inhibition

16.5   TGT: Only Functional as Homodimer

As shown in Fig. 16.1, TGT is only active as a homodimer, because one subunit accomplishes the catalysis and the second is needed to hold the tRNA in place for the reaction [11–13]. This observation motivated us to consider perturbance of the dimer formation as an alternative principle to block enzyme function. As a first step, we performed non-covalent mass spectrometry to confirm the oligomerization state of TGT and the binding stoichiometry of its complex with full-length tRNA [13]. This method enables accurate mass measurements of intact non-covalent assemblies in the gas phase. It captures the situation in solution on the protein oligomerization state as well as on protein:RNA binding stoichiometries. TGT was first analyzed alone under non-denaturing conditions. The presence of a single species was confirmed and could be assigned to dimeric TGT. NanoESI-MS experiments were then performed with TGT in the presence of increasing concentrations of tRNA. The analysis reveals that the equilibrium completely shifts towards the protein:tRNA complex and no higher order quaternary state is detected even in presence of a threefold molar excess of tRNA over TGT dimer. This finding underlines that the TGT dimer specifically binds one single tRNA molecule and corroborates the results from the above-mentioned crystallographic study showing bacterial TGT homodimer interacting with one single substrate tRNA anticodon stem loop [12]. Our results therefore definitely confirm the dimeric oligomerization state of TGT as well as the 2:1 binding stoichiometry of the TGT:tRNA complex.

16.6 Initial Mutagenesis to Disrupt Dimer Interface

Subsequently, we created two mutated variants of TGT with the aim to disrupt the dimer interface. Due to twofold symmetry of the interface, each amino acid exchange disrupts twice a certain dimer contact. In the first variant, Tyr330 was changed to Phe disrupting an H-bond of Tyr330'OH to the main chain carbonyl of Ala49. In the second variant, a salt bridge formed between the side chain ammonium group of Lys52 and the side chain carboxylate of Glu339' was eliminated by replacing Lys52 to a sterically similar but uncharged Met (Fig. 16.4).

To record the influence of the mutations on tRNA binding and catalytic efficiency, we determined $K_M(tRNA)$ and k_{cat} of both mutated variants using radiolabelled guanine as second substrate. While $K_M(tRNA)$ remained virtually unchanged for both mutated variants, k_{cat} was reduced by a factor of 10 for TGT(Tyr330Phe) and by a factor of 50 for TGT(Lys52Met). These results support the hypothesis that dimer formation is a precondition for catalytic activity of TGT. The more drastic the mutation-induced destabilization, the smaller is the fraction of catalytically active dimeric TGT. The unchanged $K_M(tRNA)$ values of both variants with respect to wild-type enzyme suggest that once the dimer has formed, it binds the tRNA substrate with virtually the same affinity as wild type TGT.

We performed nanoESI-MS experiments under non-denaturing conditions to check whether the reduced turnover of both mutated variants is due to the desta-bilized protein/protein interface. At higher concentration, TGT(Lys52Met) appears,

Fig. 16.4 Interactions within the homodimer interface of TGT (monomer A *light gray*, monomer B *dark gray*) selected for mutagenesis. The H-bond formed between Ala49 (*left*) and the Tyr330' (*right*) as well as the salt bridge formed between Lys52 (*left*) and Glu339' (*right*) are shown as *dashed lines* (distances in Å)

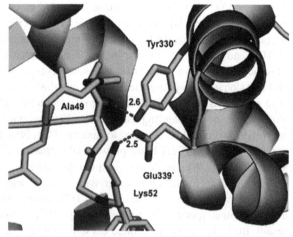

like wild type, almost exclusively as homodimer, while TGT(Tyr330Phe) also reveals a significant amount of monomer. At lower protein concentration, a substantial proportion of monomer becomes evident for both variants whereas wild type TGT remains fully dimeric at this concentration. The concentration dependence confirms that both mutations destabilize the dimer interface, well consistent with the results obtained from enzyme kinetics [13].

16.7   Computational Analysis of the Interface Stability

To study the architecture and stability of the dimer interface in a more systematical way and to better plan our next mutational experiments, we applied a computational analysis using the MM-GBSA approach [14, 15]. The stability of the C2 symmetrical homodimer interface which spans over nearly 1,600 $Å^2$ is stabilized by a patch of four aromatic amino acids (Trp326, Tyr330, His333, Phe92') and Lys52, which forms a salt bridge across the interface. These contacts are contributed twice by the two dimer mates (Fig. 16.5). As a sequence comparison across the TGT enzymes in different species shows, these residues are highly conserved [11]. The aromatic residues arrange in mutual edge-to-face stacking and achieve further stabilization by a network of hydrogen bonds using the donor functionalities of the side chains

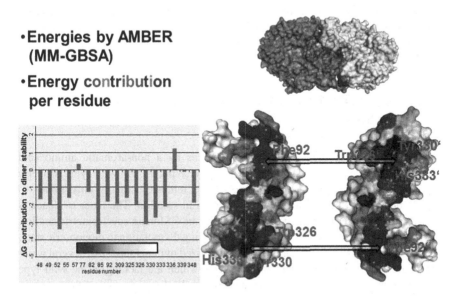

Fig. 16.5 The stability of the dimer interface was analyzed by MM-GBSA calculations. If the energy contributions (*gray* scale indicating amount of the energy contribution) are factorized on a per-residue basis, particular four aromatic residues and the salt-bridge Lys52 – Glu339 are indicated as putative stability hot spot of binding

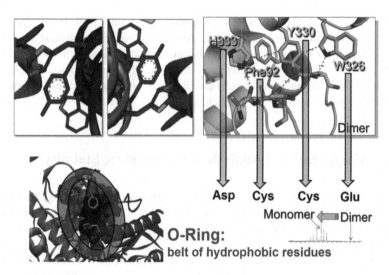

Fig. 16.6 An important motif for the stability of the dimer interface is formed by a cluster of four aromatic residues (Trp, Tyr, His, Phe). Mutation of any of these residues to a non-aromatic, more polar amino acid results in a substantial destabilization of the dimer interface. The aromatic cluster is wrapped by a ring of hydrophobic residues, which shields the cluster from water penetration and enhances electrostatic interactions

to interact with backbone carbonyl groups on the adjacent dimer mate (Fig. 16.6). This patch is embedded into a ring of hydrophobic amino acids, which supposedly shields the aromatic interaction hot spot from solvent access [16].

16.8   A Cluster of Four Aromatic Amino Acids: Essential for Stability

Mutation of any of the four aromatic residues by a non-aromatic amino acid reveals drastic loss of the homodimer stability. Whereas the wild type is nearly exclusively in dimeric state, for some of the mutated variants only minor to hardly any dimer formation could be observed in solution. The latter evidence is based on the above-mentioned non-degrading nanoESI mass spectrometric studies, which show concentration-dependent an increasing dissociation to monomeric state. Remarkably the crystal structures of all destabilized mutants show that several water molecules are incorporated into the interface, which are lacking in the wild type structure (Fig. 16.7). The mutational studies clearly show integrity of the aromatic hot-spot cluster is essential for dimer stability. Additionally, it is shielded by a ring of hydrophobic residues contributed by both monomer units of the dimer. This O ring-type motif has been suggested as important for the stability of protein-protein interfaces.

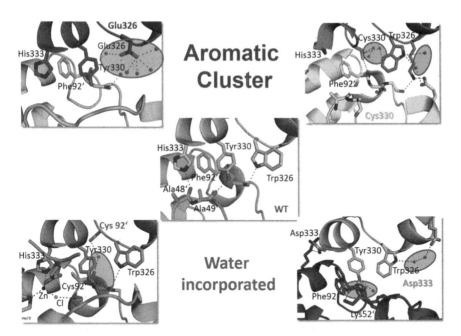

Fig. 16.7 Crystal structures of four mutant variants with altered composition of the aromatic cluster. All variants are significantly destabilized compared to the wild type (WT). In all cases, several water molecules (*encircled*) are observed that penetrate into the destabilized protein-protein interface

16.9   Switch in the Interface: a Flexible Loop-Helix Motif

Apart from the aromatic hot spot, which repeats twice due to symmetry, the interface shows an extended loop-helix motif (residues 46–62) which exhibits remarkable flexibility (Fig. 16.8). Even though this stretch comprises a fair number of polar residues, which are involved in the wild-type structure in several salt bridges and hydrogen bonds across the interface, this motif occurs in multiple conformations including a folding to helical geometry. This conformational multiplicity could be characterized in the various crystal structures determined with the wild type, its inhibitor complexes, and the mutated variants. We have hypothesized that this loop is crucial for the establishment of the interface. Supposedly, this motif is disordered in solution and adopts ordered geometry in the unperturbed wild-type interface. Most likely, it operates as a kind of shield further preventing access of water molecules to interfere and disturb the aromatic hot spots. Furthermore, it contributes to the stability by forming H-bonds across the interface via its backbone carbonyl groups to the residues of the aromatic cluster.

A special feature of this enzyme is the short distance between active site and rim of the dimer interface. This suggests, as mentioned-above, design of expanded active-site inhibitors decorated with rigid, needle-type substituents to spike into

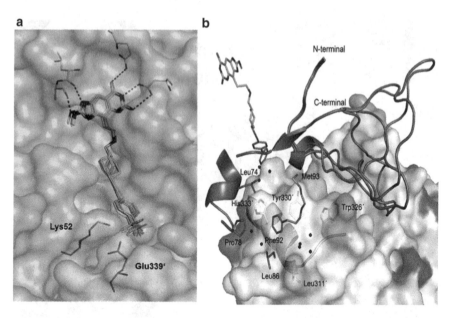

Fig. 16.8 (a) Active site inhibitors extended by long needle-like substituents penetrate into the interface region and allow to partly disrupt the interface. (b) Superposition of several structures showing the loop-helix motif in multiple conformations, which are all compatible with the dimer interface formation. In some of the determined crystal structures this entire motif is completely disordered

potential hot spots of the interaction interface [17]. Ligands with attached ethinyl-type substituents have been synthesized and characterized by K_d measurements, crystallography, non-covalent mass spectrometry, and computer simulations. In contrast to previously determined crystal structures with non-extended active-site inhibitors, the loop-helix motif, involved with well-defined geometry in several contacts across the dimer interface, falls apart and suggests enhanced flexibility once the spiking ligands are bound. Mass spectrometry indicates significant destabilization but not full disruption of the complexed TGT homodimer in solution.

In the destabilized mutant variants and in the complexes with the spiking ligands, the loop-helix motif adopts deviating conformation and orientation in the interface region (Fig. 16.8). This motivated us to envision a strategy to raise small molecule binders against this motif to morph the loop geometry in a conformation incompatible with the interface formation. A similar concept has been successfully applied in the case of small molecule modulators developed against interleukin 2 to prevent binding of this cytokine to its receptor [18].

The design of putative binders raised against a particular conformer of the loop-helix motif requires reliable information about the geometry of this motif in the monomeric state. We therefore sought for crystallization conditions to successfully crystallize the homodimeric enzyme in monomeric state. Even though some of our variants showed complete dissociation in solution (evidenced by the nanoESI MS experiments) we only succeeded in crystallizing the protein in space group C2 where the homodimeric arrangement is imposed by a crystallographic twofold axis. We broadly screened for alternative crystallization conditions, however, always ending up in the same space group with conserved dimer packing.

16.10 Artificially Introduced Disulfide Bridge Breaks Up Interface

We mutated Ile340 to cysteine, a residue located next to a putative binding pocket. Unexpectedly, mass spectrometry showed covalent linkage via a disulfide bond connecting the two monomer units to form a permanent "dimer" (Fig. 16.9a). The oxidative conditions in ambient atmosphere were sufficient to induce the observed chemical reaction and, as the Cys340 – Cys340' linkage is located next to the twofold axis, no significant change of the structural arrangement of the dimer was required.

We, therefore, applied this strategy also for other cysteine mutants. The original Tyr330Cys variant showed in solution predominantly monomeric state as indicated by the mass spectrometric studies; however in the crystal, the structure of a homodimer was determined in space group C2. Remarkably, our crystallization attempts provided additionally crystals with distinct morphology.

Diffraction data unraveled a new crystal packing in space group $P6_522$ and confirmed the expected formation of a covalent Cys330 – Cys330' disulfide linkage

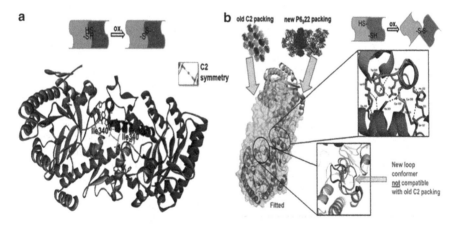

Fig. 16.9 (a) Mutation of Ile340 to Cys next to the twofold axis results in a covalently linked dimer. (b) A similar exchange of Tyr330Cys results in a covalent dimer, which enforces altered dimer packing. In the crystal a new packing in $P6_522$ instead of C2 is found. The flexible loop-helix motif adopts a new conformation (*gray*, oriented to the *left*), which would be incompatible with the original dimer interface geometry

(Fig. 16.9b). The reinforced rearrangement results in a completely different protein-protein contact. The crucial loop-helix motif, responsible to cover and interact with the aromatic hot-spot patch in the wild-type packing, is virtually exposed into "empty" space whereas the remaining monomer units are presented with virtually unchanged geometry. The loop-helix motif adopts a new, previously not yet observed conformation, which would be incompatible for steric reasons with the wild type dimer packing. The novel trace of the loop-helix motif opens a small hydrophobic pocket, which is occupied by a DMSO molecule picked-up from the cryo-buffer and a cluster of four water molecules. We could dock several candidate molecules into this pocket and we hope that such ligands will be competent to stabilize the loop-helix motif in the conformation incompatible to allow formation of the dimer interface. This would force TGT to remain in the functionally inactive monomeric state.

References

1. Klebe G (2013) Drug design, Chapter 21, Springer Reference, Heidelberg, New York, Dordrecht, London. doi:10.1007/978-3-642-17907-5
2. Romier C, Ficner R, Reuter K, Suck D (1996) Purification, crystallization, and preliminary x-ray diffraction studies of tRNA-guanine transglycosylase from *Zymomonas mobilis*. Proteins Struct Funct Genet 24:516–519
3. Grädler U, Gerber HD, Goodenough-Lashua DM, Garcia GA, Ficner R, Reuter K, Stubbs MT, Klebe G (2001) A new target for shigellosis: rational design and crystallographic studies of inhibitors of tRNA-guanine transglycosylase. J Mol Biol 306:455–467
4. Stengl B, Reuter K, Klebe G (2005) Mechanism and substrate specificity of tRNA – guanine transglycosylases (TGTs): tRNA modifying enzymes from thee three different kingdoms of life seem to share a common mechanism. ChemBioChem 6:1–15
5. Brenk R, Naerum L, Grädler U, Gerber HD, Garcia GA, Reuter K, Stubbs MT, Klebe G (2003) Virtual screening for submicromolar leads of TGT based on a new unexpected binding mode detected by crystal structure analysis. J Med Chem 46:1133–1143
6. Meyer EA, Donati N, Guillot M, Schweizer B, Diederich F, Stengl B, Brenk R, Reuter K, Klebe G (2006) Synthesis, biological evaluation, and crystallographic studies of extended guanine-based (lin-benzoguanine) inhibitors for tRNA-guanine transglycosylase (TGT). Helv Chim Acta 89:573–597
7. Hörtner S, Ritschel T, Stengl B, Kramer C, Klebe G, Diederich F (2007) Potent inhibitors of tRNA-guanine transglycosylase, an enzyme linked to the pathogenicity of the Shigella bacterium: charge-assisted hydrogen bonding. Angew Chem Int Ed 46:8266–8269
8. Ritschel T, Kohler PC, Neudert G, Heine A, Diederich F, Klebe G (2009) How to replace the residual solvation shell of polar active site residues to achieve nanomolar inhibition of tRNA-guanine transglycosylase. ChemMedChem 4:2012–2023
9. Kohler PC, Ritschel T, Schweizer WB, Klebe G, Diederich F (2009) High-affinity inhibitors of tRNA-guanine transglycosylase replacing the function of a structural water cluster. Chem Eur J 15:10809–10817
10. BarandunL IF, Kohler PC, Ritschel T, Heine A, Orlando P, Klebe G, Diederich F (2013) High-affinity inhibitors of Zymomonas mobilis tRNA–guanine transglycosylase through convergent optimization. Acta Crystallogr D69:1798–1807

11. Stengl B, Meyer EA, Heine A, Brenk R, Diederich F, Klebe G (2007) Crystal structures of tRNA-guanine transglycosylase (TGT) in complex with novel and potent inhibitors unravel pronounced induced-fit adaptations and suggest dimer formation upon substrate binding. J Mol Biol 370:492–511
12. Xie W, Liu X, Huang RH (2003) Chemical trapping and crystal structure of a catalytic tRNA guanine transglycosylase covalent intermediate. Nat Struct Biol 10:781–788
13. Ritschel T, Atmanene C, Reuter K, Van Dorsselaer A, Sanglier-Cianferani S, Klebe G (2009) An integrative approach combining noncovalent mass spectrometry, enzyme kinetics and X-ray crystallography to decipher Tgt protein-protein and protein-RNA interaction. J Mol Biol 393:833–847
14. Gohlke H, Kiel C, Case DA (2003) Insights into protein-protein binding by binding free energy calculation and free energy decomposition for the Ras-Raf and Ras-RalGDS complexes. J Mol Biol 330:891–913
15. Jakobi S, Nguyen TXP, Debaene F, Metz A, Sanglier-Cianferani S, Reuter K, Klebe G (2014) Hot-spot analysis to dissect the functional protein–protein interface of a tRNA-modifying enzyme. Proteins Struct Funct Bioinform 82:2713–32
16. Bogan AA, Thorn KS (1998) Anatomy of hot spots in protein interfaces. J Mol Biol 280:1–9
17. Immekus F, Barandun LJ, Betz M, Debaene F, Petiot S, Sanglier-Cianferani S, Reuter K, Diederich F, Klebe G (2013) Launching spiking ligands into a protein–protein interface: a promising strategy to destabilize and break interface formation in a tRNA modifying enzyme. ACS Chem Biol 8:1163–1178
18. Arkin MR, Wells JA (2004) Small-molecule inhibitors of protein–protein interactions: progressing towards the dream. Nat Rev Drug Discov 3:301–317

Chapter 17
Molecular Interaction Analysis for Discovery of Drugs Targeting Enzymes and for Resolving Biological Function

U. Helena Danielson

Abstract Analysis of molecular interactions using surface plasmon resonance (SPR) biosensor technology has become a powerful tool for discovery of drugs targeting enzymes and resolving biological function. A major advantage of this technology over other methods for interaction analysis is that it can provide the kinetic details of interactions. This is a consequence of the time resolution of the analysis, which allows individual kinetic rate constants as well as affinities to be determined. A less commonly recognized feature of this technology is that it can reveal the characteristics of more complex mechanisms, e.g. involving multiple steps or conformations of the target or ligand, as well as the energetics, thermodynamics and forces involved.

17.1 Introduction

17.1.1 SPR Biosensor Basics

The application of SPR biosensor technology for drug discovery is now well recognized and has been reviewed previously, e.g. [1–4]. However, due to the rapid development of the technology and experimental procedures, the full potential of what can actually be achieved today is not widely known. In addition, there are some misconceptions remaining from the early days of SPR biosensors, which make many researchers unaware of the possibilities or new users hesitant to take on the technology for their problems of interest. The unawareness primarily concerns the high sensitivity of measurements and the broad repertoire of experiments that can be performed, contributing to the exceptionally high information content of data. In order to put the application of SPR biosensor analysis into perspective and to encourage more researchers to use the technology, both for low molecular weight

U.H. Danielson (✉)
Department of Chemistry - BMC, Uppsala University, Uppsala, Sweden
e-mail: helena.danielson@kemi.uu.se

© Springer Science+Business Media Dordrecht 2015   223
G. Scapin et al. (eds.), *Multifaceted Roles of Crystallography in Modern Drug Discovery*, NATO Science for Peace and Security Series A: Chemistry and Biology, DOI 10.1007/978-94-017-9719-1_17

drug discovery and elucidation of biological function, a brief description of SPR biosensor technology and the information that can be obtained is justified.

This brief technical background is focused on the methodology as implemented in the range of Biacore instruments designed for the analysis of low molecular weight analytes, the first commercialized SPR biosensor. For a more detailed description of the technical features of these instruments, as well as the unique principles for experimental design and data analysis, see the Biacore handbook [5]. However, there are today also other SPR-based technologies that have adequate sensitivity, time resolution and flexibility in experimental design for small molecule drug discovery. The performance and applicability of these methods is not as well described since the number of published studies using these technologies is rather limited so far. However, a benchmark study of affinity-based biosensors gives a good overview of the systems commonly used today [6].

17.1.2 Detection of Interactions Using SPR Biosensors

SPR biosensors can be seen as flat affinity chromatography surfaces that in real time sense the interaction between immobilized molecules and analytes injected into the mobile phase (Fig. 17.1). SPR is an optical phenomenon that is associated with the total internal reflection of light at the boundary between two media of different optical properties, described by their different dielectric functions [7]. In the SPR biosensor, plane polarized light is reflected by a thin gold film on a sensor chip that is positioned in a microfluidic system [8, 9]. An advantage is that it does not require labeling of any of the interactants or reporter molecules. The light never passes through the sample so there is no absorption or dispersion of light. The signal is influenced by the small changes in refractive index that molecules close to the surface can induce and is recorded in refractive (or resonance) units (RU).

A potential disadvantage is the need to immobilize one of the binding partners to the sensor matrix. However, there are multiple strategies for immobilization, and

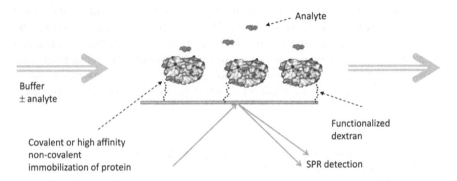

Fig. 17.1 Principles for SPR biosensor technology for analysis of molecular interactions

the dextran matrix to which the sensing molecule is attached provides a favorable environment for interaction analysis. For proteins, this has been found to have advantages over methods where the proteins are free in solution, which often require higher concentrations and can therefore lead to aggregation.

Interactions between low molecular weight analytes and immobilized proteins are often thought to be undetectable since there is a correlation between the magnitude of the signal and the change in mass upon the binding of an analyte to the surface. But even though the change in refractive index at the surface is associated with the mass of the analyte, current instruments have the sensitivity required for analysis even of small molecules. However, there are other effects that influence the refractive index and that may be significant in experiments involving low molecular weight analytes [10]. The advantage of the general detection principle may thus become a challenge when the signal is to be interpreted mechanistically. SPR biosensor-based experiments therefore require appropriate and extensive references and controls, as well as cautious mechanistic interpretation of results, especially for more complex interactions.

17.1.3 Information Content

For simplicity, it is assumed in this text that the immobilized molecule is a protein while the analyte is a small organic ligand or another protein. The simplest interaction assumed is the reversible 1:1 interaction, which is described by the following schemes and equations:

$$P + L \rightleftharpoons PL$$

$$K_D = k_{off} / k_{on} = [P] [L] / [PL]$$

$$[PL] = [P_{tot}] * [L] / K_D * [L]$$

The information that can be extracted from interactions that result in detectable signals is consequently related to this formal description. However, the experimental design and the type of data analysis used influences how the information can be obtained.

Figure 17.2 illustrates the difference between analysis of complete progress curves for interactions (left graph) and the commonly used graph for equilibrium-based data of complex formation as a function of ligand concentration (right graph). Report points extracted during the steady state phase in the progress curves are used to generate the saturation curve. A good understanding of the mathematical principles for data analysis and the equations to be used is required for extraction of meaningful information from biosensor experiments. Much of the analysis is taken care of by the software accompanying the instrument.

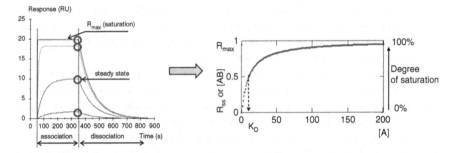

Fig. 17.2 Relationship between progress curves for association and dissociation of analytes interacting with a sensor surface (sensorgrams) and steady state signals (O) as a function of analyte concentration. The graphs represent simulated data for a reversible 1:1 interaction (Color figure online)

There are three principal levels of analysis:

Low resolution analysis: Report point-based data extracted after one or several defined times after injection, not necessarily at steady state. By comparing the signal for test compounds with those for positive and negative references it is possible to conclude if an analyte interacts with the immobilized protein or not. It is an efficient method that is often sufficient for screening and pilot studies.

Medium resolution analysis: Analysis of equilibrium data extracted from the steady state signals of progress curves obtained with a series of different analyte concentrations (Fig. 17.2, left). The data can be used to estimate the basic interaction mechanism and equilibrium parameters, i.e. stoichiometry and affinity (K_D). It is often used when kinetic rate constants are too fast or slow to be quantified.

High resolution analysis: Global analysis of progress curve data representing the complete time course for association and dissociation, ideally for several concentrations of analyte (Fig. 17.2, left graph). This type of data can be used to define the mechanism of the interaction and determine the corresponding kinetic rate constants. In order to fully exploit the information potential of SPR biosensor experiments, the more extensive experimental designs for high resolution analysis are required.

An additional type of information that can be obtained is related to experimental artifacts seen as atypical sensorgrams that arise as a consequence of aggregation, micelle formation or precipitation. In contrast to many other methods, these are often seen directly in the raw data, especially at high concentrations of analyte [11]. This information can be very important, especially for identifying false positives and understanding the reason for false negatives. However, in order to reduce the risk of misinterpretations, compounds are typically analysed in 5 % DMSO at concentrations below those where any problems with solubility are expected. Concentrations can often be up to 500 μM or even 1 mM, but many compounds can only be studied at much lower concentrations.

17.1.4   Kinetic and Affinity Data

Kinetic analysis of molecular interactions is clearly the most common application for SPR biosensors. The range of rate constants and affinities that can be determined has been well described and the methodological aspects considered [12]. In principle, the rate of diffusion and the ability to distinguish an irreversible interaction from one with a very slow dissociation sets the limit for the rate constants that can be determined. Affinities can typically be determined if it is possible to study the interactions with analyte concentrations that are significantly higher than the K_D-value. This therefore depends on the solubility of the analyte. For interactions that are more complex than simple 1:1 interactions, e.g. involving multiple binding sites or conformational changes, the requirement for high-quality data over a wide range of concentrations (including concentrations $\gg K_D{}^{app}$) alters the limits of values and influences which parameters can be reliably quantified. Moreover, the availability of references (i.e. both reference compounds and reference proteins) is essential for validation of results and determines the reliability of the interpretations. More complex data require a realistic theoretical model that can be confirmed by other types of experiments or computational modelling.

Extensive data analysis using a well-defined mechanistic model and a quantitative determination of kinetic parameters is not always required as sensorgrams provide a very simple and easily interpreted qualitative analysis of the kinetics of interactions. This is often sufficient for comparative studies of lead series or target selectivity. For example, the effect of analyte concentrations and dissociation rates on both complex concentration and stability over time is illustrated in Fig. 17.3. This

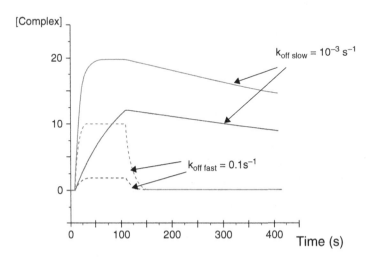

Fig. 17.3 Simulated sensorgrams for interactions with the same k_{on} (10^6 M^{-1} s^{-1}) but different k_{off} for two different analyte concentrations (1,000 nM (*top, blue*), or 100 nM (*bottom, red*)) and the same target concentration. The corresponding affinities can be calculated from: $K_D = k_{off}/k_{on}$ (Color figure online)

is clearly more informative than equilibrium-based measurements, which simply detect the differences in steady-state levels, but without providing the underlying explanation for the results.

The importance of slow dissociation of drug-target complexes has recently emerged as a primary kinetic characteristic of importance for lead optimization (cf. [13, 14]). Although this is not a new concept as such, being one of the key parameters determined in pharmacokinetic studies. But using a biosensor-based approach it is now possible to detect compounds with certain kinetic features already at the level of screening [15] and dissociation rates (also expressed as residence time, $\tau = 1/k_{off}$) can be determined during the characterization of hits and leads, which makes it an attractive parameter to use also in the early stages of lead discovery. Despite the obvious advantage of drugs with slow dissociation, there are instances where other parameters are more important, e.g. as in the case of HIV protease inhibitors [16]. Also, if slow dissociation was the ultimate goal for optimization, irreversible inhibitors would be the ideal solution. Although many drugs today are irreversible, it is a feature that may cause side effects and is therefore a characteristic often avoided. It is therefore essential to establish the kinetic parameter(s) and other interaction kinetic features that correlate with good *in vivo* efficacy before optimizing a lead on the basis of a single kinetic parameter.

17.1.5   Mechanistic Analysis

Molecular interactions are typically assumed to be simple and reversible 1:1 interactions, without additional molecular species or steps due to secondary effects such as conformational changes or chemical transitions. Although this may often be the case, more complex mechanisms are common. The mechanism should consequently be verified for each target and type of compound.

Important basic mechanistic information about an interaction is readily obtained since reversibility, stoichiometry and the number of steps involved in forming a complex are observed directly as deviations from ideal behaviour in the data – providing that experiments are well designed. Mechanistic complexities are therefore often detected already when an assay for a target is established since the sensorgrams are not well described by a simple 1:1 interaction model. They are not as easily recognized when using equilibrium based interaction assays or indirect assays, such as enzyme inhibition assays. For example, detection of enzyme inactivation or time dependent inhibition with a substrate-based inhibition assay requires a specifically designed experiment. Even more elaborate experimental designs are required to establish the mechanism for the time dependence [17]. The researcher must therefore suspect that there is a complexity that should be followed up. Although mechanistic complexities are readily detected with SPR biosensor experiments, it is not trivial to determine the kinetic rate constants for the individual steps. For example, the quantitative analysis of tight-binding inhibitors can be challenging if there are no suitable regeneration conditions available. Experiments may therefore be limited to qualitative analysis using single injections over a sensor surface [15, 18] or "single cycle kinetics" [19]. However, qualitative analysis of an interaction for which the mechanism is known to be complex is often more informative than the blind use of equilibrium based inhibition parameters (e.g. K_i or IC_{50}-values), which can be very misleading.

17.2   Proof-of-Concept Studies Using HIV-1 Protease

The earliest study on the application of SPR biosensor analysis for kinetic character-ization of lead compounds was performed with HIV-1 protease as a target [20]. The technology had previously required that the analyte was macromolecular, but the newly launched SPR biosensor instrument enabled the use of a reference surface so that non-specific signals could be subtracted from the target protein surface. This is essential for the specific detection of interaction between low molecular weight analytes and protein surfaces. The method was found very reliable as it also revealed artifacts due to nonspecific signals, incomplete regeneration, and carryover. Although the original experiments involved the use of report points and a simple graphical display of data, they revealed that the time resolution was sufficient for kinetic characterization of the interactions. Subsequent studies improved the experimental design and analysis procedures, enabling more advanced studies of lead compounds. Nevertheless, the initial experiments showed an important proof-of-principle and demonstrated that it was possible to distinguish HIV-1 inhibitors from other compounds in a randomized series and that they differed in their interaction kinetics.

17.2.1   Structure-Kinetic Relationship Analysis

An advantage of time-resolved methods for lead characterization is the possibility of profiling of series of lead compounds via structure-kinetic relationship analysis (SKR), adding two dimensions to the more conventional structure-activity relation-ship (SAR) analysis [21]. The advantage is immediately evident by visualizing the kinetic profiles of compounds in an interaction kinetic plot (Fig. 17.4). It reveals that compounds that have similar affinities (on the same diagonal) may differ significantly in their kinetics. This is relevant since the optimization of a compound interaction with slow kinetics must be accomplished by different structural changes than the optimization of a compound interaction with fast kinetics. How this can be done in practice is still obscure, at least on a general level.

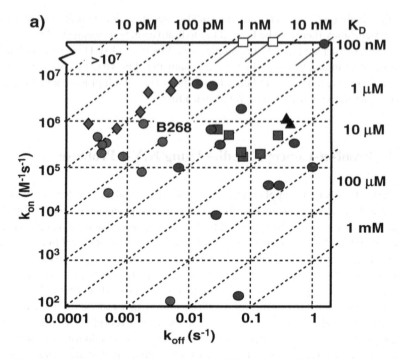

Fig. 17.4 Structure-kinetic relationship analysis of HIV-1 protease inhibitors, data from [21]. Drugs and related compounds (♦), P1/P1′ analogues of B268 (●), P2/P2′ analogues of B268 (●), cyclic ureas (□) and cyclic sulfamides (■), and non-B268 analogues (▲) (Color figure online)

17.2.2 Selectivity Analysis and Resistance Profiling

The quantification and visualization of structure-kinetic relationships has been demonstrated to be very powerful also for assessing selectivity. An example is the analysis of the effect of one, two or four amino acid substitutions in HIV-1 protease on the kinetics of interactions with inhibitors (Fig. 17.5). The graph shows the resistance profile of the inhibitors and that the reduced affinity for the resistant enzyme variants correlated with both slower association and faster dissociation rates. The profiling contributes to the understanding of how mutations in the viral genome influences the efficacy of inhibitors and how medicinal chemists can optimize compounds for a higher resilience to resistance development.

A type of selectivity study and the search for broad spectrum drugs involved the screening of a small library of HIV-1 protease inhibitors against three isoenzymes of secreted aspartic proteases (SAPs 1–3) from *Candida albicans* [23]. Important differences in the selectivity of these types of compounds for the different isoenzymes were identified and enabled the identification of inhibitors that had relatively

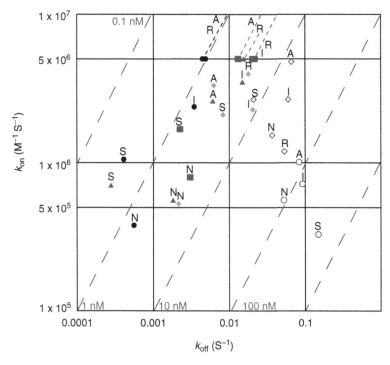

Fig. 17.5 Resistance profiling of HIV-1 protease inhibitors [22]. The data points are labelled by the name of the compound: *S* saquinavir, *I* indinavir, *N* nelfinavir, *A* amprenavir, *R* ritonavir, and the symbols represent the enzyme variants: wild-type (●), L90M (■), G48V (♦), V82A (▲), I84V/L90M (◊), and G48V/V82A/I84V/L90M (○) (Color figure online)

high affinity for all three isoenzymes. This suggested that broad-spectrum inhibitors of SAPs may be designed and that ritonavir was an example of such a compound.

Similarly, hepatitis C virus (HCV) protease inhibitors have been profiled against three different genotypes (1a, 1b and 3a) of the target protein [24]. The different selectivity profiles of the inhibitors reveal the potential of the compounds to act as broad spectrum inhibitors. This is relevant since none of the clinical drugs and those in advanced clinical trials inhibit all clinically relevant genotypes adequately. Selectivity analysis can therefore contribute to the identification and optimization of next generation anti-HCV drugs.

Due to the large number of proteases naturally occurring in nature and variants that evolve as a result of drug pressure the question of selectivity needs to be addressed. These examples show that the high resolution of SPR-based interaction kinetic studies enables in-depth analysis of features related to selectivity that are difficult to obtain by more conventional methods (see also below).

17.2.3   Time-Resolved Thermodynamic Analysis

The principles and the type of thermodynamic data that can be generated by SPR biosensor technology were initially demonstrated for HIV-1 protease and a series of inhibitors [25]. Although the data is relatively simple to obtain, the molecular interpretation and relevance for drug discovery remains to be demonstrated. A more in-depth study supported by stopped-flow, ITC and structural data has therefore more recently been performed with thrombin and melagatran analogues [26].

Conventional thermodynamic analysis involves direct measurements of heat changes upon ligand binding to a target protein at equilibrium using isothermal calorimetry (ITC). The data can be used to calculate changes in enthalpy, entropy and Gibbs free energy ($\Delta G = \Delta H - T \cdot \Delta S$), the latter easily converted to affinity ($\Delta G = - R \cdot T \cdot \ln K_D$). This can provide important insights into the thermodynamic driving forces of an interaction, but pharmaceutical researchers often use the method simply for determination of stoichiometries and K_D-values. For a more detailed interpretation it is essential to have access to high-resolution structural data, which has been demonstrated, for example, in a well-designed study of the thermodynamics of interactions between thrombin and a series of inhibitors [27].

However, the thermodynamics of interactions can also be indirectly determined by time resolved methods. This is achieved by measuring the kinetics of an interaction over a range of temperatures and then using Arrhenius and Eyring analysis for estimation of the thermodynamic parameters for association and dissociation. Figure 17.6 illustrates that both association and dissociation rates increase with temperature, but that the correlation is slightly different for different compounds. These differences translate into different thermodynamic profiles for the compounds.

The study involving thrombin and melagatran analogues contributed to important insights into structure-thermodynamic relationships, for example that hydrogen bond formation and breakage is not necessarily reflected in enthalpy gains and losses, respectively, which is in contrast to what may be intuitively expected. However, the relevance of thermodynamic data for drug design will require a larger set of inhibitor-target systems for a broad analysis of the chemical principles and how they relate to biological readouts, such as efficacy and specificity, for example.

Despite the interest initially spurred by the idea that it is an advantage to select hits and leads that have a dominant enthalpic component in order to avoid an optimization process that results in compounds with a too large entropic contribution, the thermodynamic profiling of lead compounds has not really had the impact on the drug discovery process that was originally expected [28]. The difficulties in implementing this approach can most likely be attributed to the complexities of the system, resulting from multiple components and their independent and dependent conformational changes, as well as the often large impact of solvent. Interpretations of a certain thermodynamic profile a molecular level is consequently very difficult and the few detailed studies published do not support the simplistic interpretations often proposed.

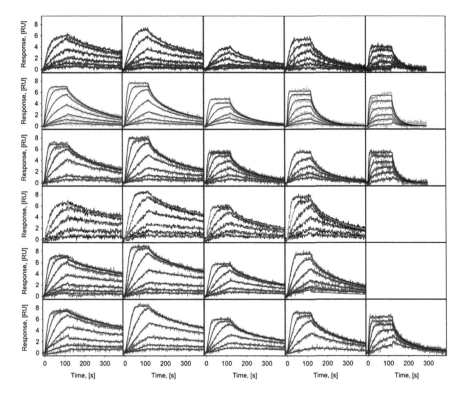

Fig. 17.6 Effect of temperature on the kinetics of interactions between thrombin and melagatran (*top*) and a series of melagatran analogues with the hydroxyl group in the P3 carboxyl group replaced with amine residues of increasing chain length and hydrophobicity (from *top* to *bottom*) at 5, 15, 25, 35, and 45 °C (from *left* to *right*). The maximal signal responses were not normalized and therefore vary for different datasets, this does not influence the kinetic profiles (Data from [26])

17.2.4 Chemodynamics

"Chemodynamics" is a term we have used to describe time-resolved relationships between ligand-target interactions and the chemical environment, such as pH or ionic strength. This is analogous to the temperature dependence of interactions used for thermodynamic analysis (see above). Although most chemists will intuitively understand that chemical interactions are dependent on both temperature and the chemical environment, it is rather surprising that few studies are designed to use this phenomenon to extract information about the system of interest. Instead, much effort is focused on keeping the environment constant and as close as possible to a perceived physiological situation. The latter is probably rather naïve since a biological system is considerably more complex than the *in vitro* systems used for biochemical studies and matching environments are simply not possible due to crowding effects and local high concentrations.

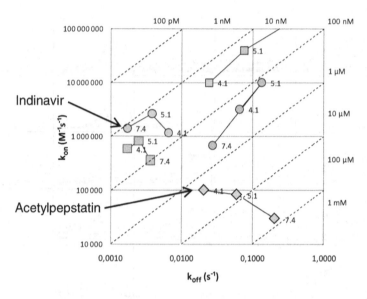

Fig. 17.7 pH-dependency profile of HIV-1 protease inhibitors (Data from [29])

An initial chemodynamic study revealed that the pH dependency of HIV-1 protease inhibitors was a unique kinetic characteristic of the compound [29], illustrated in Fig. 17.7. It can be noted that association and dissociation had different pH-profiles, indicating that the two processes proceeded by different pathways/mechanisms. Similar results have been obtained also for secreted aspartic acid proteases (SAPs) from *Candida albicans* [30], revealing that there were significant differences between SAP1 and SAP2 for acetyl-pepstatin, showing that also differences in the target will influence the pH profiles.

A rather simplistic structural analysis of the interactions with HIV-1 proteases suggested that the profile for indinavir could be explained by interactions with a non-catalytic aspartate and an arginine residue (Fig. 17.8).

Later work has involved more sophisticated modelling and focused more on the protonation states of the aspartic acid residues, which are affected by the binding of the inhibitors [31, 32]. The large number of additional ionisable groups in the enzyme and often also inhibitors makes the analysis rather challenging.

Although the modelling so far has been based on affinities, it should be possible to perform corresponding modelling on the basis of the kinetic rate constants. But this awaits advances in modelling procedures – the data is available for any researchers willing to take on the challenge. Nevertheless, even if all the information cannot yet be used for modelling, and thus for prediction and design, this type of analysis makes it possible to identify the dominating interaction forces in association and dissociation separately.

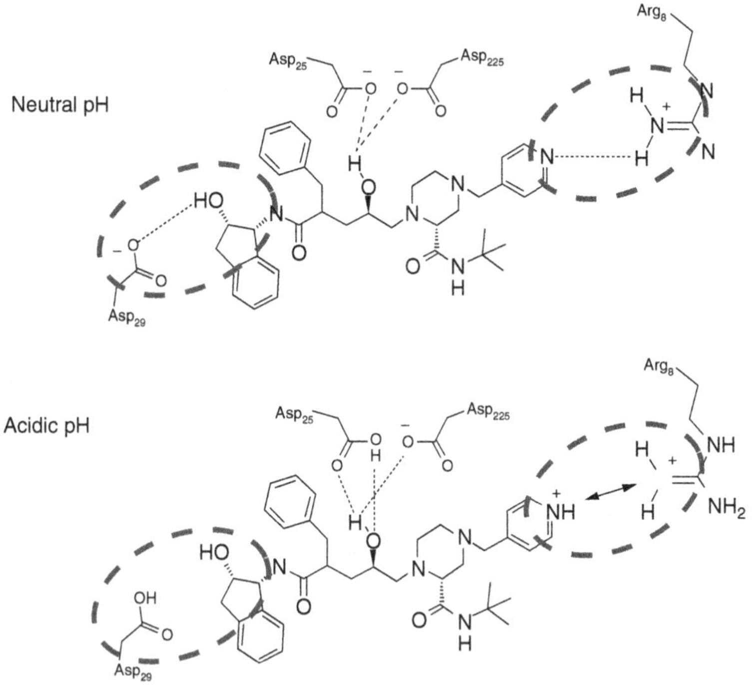

Fig. 17.8 Indinavir interaction with HIV-1 protease at different pH [29]

17.3   Mechanistic Studies – From Simple to Complex Interactions

The kinetic analysis of interactions between lead compounds and their targets has been found to be very useful, as illustrated above. But molecular interactions are often more complex than that described by a reversible 1-step interaction, which has been assumed in all the cases presented so far. SPR biosensors are well suited to both detecting mechanistic complexities and characterizing them.

17.3.1   Irreversible vs. Slow Dissociation

A common cause for difficulties when analysing lead compounds is irreversibility, either due to a very tight interaction or the formation of a stable covalent bond. This is rather common once leads have gone through an optimization process. However, although such compounds are perceived as "good", it can be difficult to discriminate and rank series of compounds since the slow dissociation will prevent other features from being characterized. Conventional equilibrium-based methods, such as steady state-enzyme kinetic measurements, can be used to generate K_i or

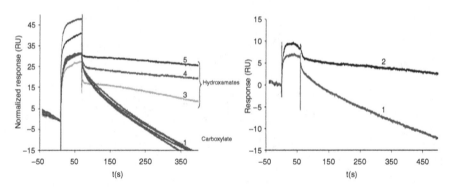

Fig. 17.9 Comparison of kinetic profiles of MMP-12 inhibitors with a hydroxamate group (2–5) and an analogue with a carboxylate group (1)

IC_{50} values. However, this is only possible up to a point where the interaction can be reliably assumed to be measured at equilibrium, which is not always evident or experimentally verified. Interactions with slower dissociation can be monitored by progress curve-based analyses, which can also provide inactivation rate constants. However, these are suboptimal since they rely on an indirect substrate-based readout and experimental systems that are stable for relatively long measurement times. A direct time-resolved assay therefore has an exceptional capacity to provide informative data also for irreversible and very slow interactions. However, for this method, the accuracy of the analysis is dependent on the stability of the target surface. But by using suitable references it is possible to evaluate the functionality of the surface over time.

Using an SPR biosensor-based approach it was revealed that MMP-12 inhibitors containing the popular hydroxamate group interact essentially irreversibly with their target [33]. Figure 17.9 illustrates that a carboxylic acid analogue dissociates faster than the hydroxamates, confirming that the tight binding can be attributed to chelation with the catalytic zinc of MMPs. The slow dissociation of this interaction is advantageous for potency, but is problematic from a selectivity perspective since it can be argued that binding, dissociation and rebinding is an amplification process that increases the desired population of complexes. Although the aim has been to introduce other groups into hydroxamate inhibitors to achieve specificity, this has been found to be very challenging, as demonstrated by the fact that a relatively large number of potent hydroxamate inhibitors have failed in the clinic due to toxic effects.

17.3.2 Mechanism-Based Inhibition

The possibility of designing selective and efficient drugs by using mechanism-based inhibition of proteases is well established. It is a strategy primarily used for serine and thiol proteases since both of these classes of proteases catalyse hydrolysis via a two-step mechanism involving a covalent acyl enzyme intermediate. This means that two populations of protein-ligand complex exist; the initially formed encounter complex and a more stable complex held together by a reversible covalent bond. However, the relative amounts of the two complexes and the contribution of the covalent bond to the stability of the complex are not known a priori since they depend on the four rate constants and the concentrations of the free ligand and protein. The advantage of a mechanism-based mode-of-action may therefore not be significant [34]. However, by using a kinetic approach the details of the interaction and inhibition mechanism can be readily established.

The non-structural protein 3 (NS3) of hepatitis C virus (HCV) harbours a serine protease in its N-terminal domain. It has been considered to be a suitable target for drugs and three protease inhibitors, telaprevir, boceprevir and simeprevir, are now available for treatment of HCV infections. Telaprevir and boceprevir are both mechanism-based and use an electrophilic warhead to form a reversible covalent bond with the catalytic serine residue [35, 36]. Although these compounds are very important complements to previous therapy, they still need to be given in combination with interferon α and ribavirin. Moreover, they are not effective against all relevant strains of the virus and resistance will be an emerging problem, similar to the situation for HIV protease inhibitors [36].

To aid in the design of next generation HCV inhibitors the interaction mechanisms and kinetics of these compounds have been analysed the characteristics compared with other compounds that have reached clinical trials [37] (and work submitted for publication). The two compounds both interact with a mechanism involving two steps and that result in the formation of a covalent complex with the target. Kinetic studies have confirmed the formation of this complex (Fig. 17.10), demonstrating that this approach is useful.

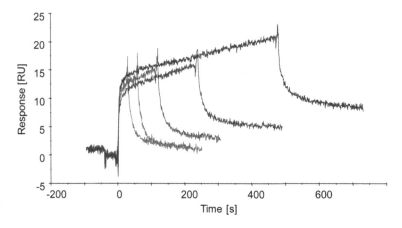

Fig. 17.10 Comparison of sensorgrams for the same concentration of telaprevir injected over immobilized HCV NS3 protease for different times [37]

However, due to other structural features, the kinetics of the compounds are suboptimal. In comparison to two macrocyclic inhibitor BILN-2061 [38] and ITMN-191 [39] the association rates were comparatively slow and the dissociation rates fast, resulting in a relatively low affinity. For both boceprevir and telaprevir the two-step mechanism was therefore essential for efficacy. The two analysed macrocyclic compounds clearly had favourable kinetics although BILN-2061 has other properties that make the compound unsuitable as a drug. The clinical usefulness of ITMN-191 is still being explored.

17.4 Biological Function

The use of SPR biosensors for analysis of biological function has been demonstrated with a number of molecular systems. An illustrative example is where the interaction between caldendrin and AKAP79 is compared to that for calmodulin and AKAP79, and their different dependence on calcium ions [40]. The data suggests that the two proteins have different roles in synaptic function.

Another example where the method has provided new insight into biological function is that of $GABA_A$-receptors [41]. This ligand gated ion channel was demonstrated to interact with histamine at affinities in the micromolar range, suggesting that it may be a histaminergic receptor. This is expected to provide new insights into histaminergic pharmacology and provide new strategies for discovery of drugs interfering with cys-loop receptors, an important group of receptors in the brain.

References

1. Huberand W, Mueller F (2006) Biomolecular interaction analysis in drug discovery using surface plasmon resonance technology. Curr Pharm Des 12(31):3999–4021
2. Pröll F, Fechner P, Pröll G (2009) Direct optical detection in fragment-based screening. Anal Bioanal Chem 393(6–7):1557–1562
3. Danielson UH (2009) Integrating surface plasmon resonance biosensor-based interaction kinetic analyses into the lead discovery and optimization process. Future Med Chem 1(8):1399–1414
4. Danielson UH (2009) Fragment library screening and lead characterization using SPR biosensors. Curr Top Med Chem 9(18):1725–1735
5. Healthcare GE (2012) Biacore assay handbook. Bio-Sciences AB, Uppsala, Sweden
6. Rich RL et al (2009) A global benchmark study using affinity-based biosensors. Anal Biochem 386(2):194–216
7. Knoll W (1998) Interfaces and thin films as seen by bound electromagnetic waves. Annu Rev Phys Chem 49:569–638
8. Stenberg E et al (1991) Quantitative determination of surface concentration of protein with surface plasmon resonance using radiolabeled proteins. J Colloid Interface Sci 143(2):513–526
9. Malmqvist M (1993) Biospecific interaction analysis using biosensor technology. Nature 361(6408):186–187

10. Davis TM, Wilson WD (2000) Determination of the refractive index increments of small molecules for correction of surface plasmon resonance data. Anal Biochem 284(2):348–353
11. Giannetti AM, Koch BD, Browner MF (2008) Surface plasmon resonance based assay for the detection and characterization of promiscuous inhibitors. J Med Chem 51(3):574–580
12. Önell A, Andersson K (2005) Kinetic determinations of molecular interactions using Biacore– minimum data requirements for efficient experimental design. J Mol Recognit 18(4):307–317
13. Copeland RA, Pompliano DL, Meek TD (2006) Drug-target residence time and its implications for lead optimization. Nat Rev Drug Discov 5(9):730–7399
14. Swinney DC (2009) The role of binding kinetics in therapeutically useful drug action. Curr Opin Drug Discov Dev 12(1):31–39
15. Elinder M et al (2009) Screening for NNRTIs with slow dissociation and high affinity for a panel of HIV-1 RT variants. J Biomol Screen 14(4):395–403
16. Shuman CF, Vrang L, Danielson UH (2004) Improved structure-activity relationship analysis of HIV-1 protease inhibitors using interaction kinetic data. J Med Chem 47(24):5953–5961
17. Copeland RA (2005) Evaluation of enzyme inhibitors in drug discovery. Wiley, Hoboken
18. Elinder M et al (2010) Inhibition of resistant HIV-1 by MIV-170, a slowly dissociating non-nucleoside reverse transcriptase inhibitor. Biochem Pharmacol 80(8):1133–1140
19. Radi M et al (2009) Discovery of chiral cyclopropyl dihydro-alkylthio-benzyl-oxopyrimidine (S-DABO) derivatives as potent HIV-1 reverse transcriptase inhibitors with high activity against clinically relevant mutants. J Med Chem 52(3):840–851
20. Markgren PO, Hämäläinen M (1998) Danielson UH (1998) Screening of compounds interacting with HIV-1 proteinase using optical biosensor technology. Anal Biochem 265(2):340–350
21. Markgren PO et al (2002) Relationships between structure and interaction kinetics for HIV-1 protease inhibitors. J Med Chem 45(25):5430–5439
22. Shuman CF et al (2003) Elucidation of HIV-1 protease resistance by characterization of interaction kinetics between inhibitors and enzyme variants. Antivir Res 58(3):235–242
23. Backman D, Monod M, Danielson UH (2006) Biosensor-based screening and characterization of HIV-1 inhibitor interactions with Sap 1, Sap 2, and Sap 3 from Candida albicans. J Biomol Screen 11(2):165–175
24. Ehrenberg AE et al (2014) Accounting for strain variations and resistance mutations in the characterization of hepatitis C NS3 protease inhibitors. J Enzyme Inhib Med Chem 29:868–876
25. Shuman CF, Hämäläinen MD, Danielson UH (2004) Kinetic and thermodynamic characterization of HIV-1 protease inhibitors. J Mol Recognit 17(2):106–119
26. Winquist J et al (2013) Identification of structural-kinetic and structural-thermodynamic relationships for thrombin inhibitors. Biochemistry 52(4):613–626
27. Biela A et al (2012) Ligand binding stepwise disrupts water network in thrombin: enthalpic and entropic changes reveal classical hydrophobic effect. J Med Chem 55(13):6094–6110
28. Freire E (2008) Do enthalpy and entropy distinguish first in class from best in class? Drug Discov Today 13(19–20):869–874
29. Gossas T, Danielson UH (2003) Analysis of the pH-dependencies of the association and dissociation kinetics of HIV-1 protease inhibitors. J Mol Recognit 16(4):203–212
30. Backman D, Danielson UH (2003) Kinetic and mechanistic analysis of the association and dissociation of inhibitors interacting with secreted aspartic acid proteases 1 and 2 from Candida albicans. Biochim Biophys Acta 1646(1–2):184–195
31. Dominguez JL et al (2012) Experimental and 'in silico' analysis of the effect of pH on HIV-1 protease inhibitor affinity: Implications for the charge state of the protein ionogenic groups. Bioorg Med Chem 20(15):4838–4847
32. Sussman F et al (2012) On the active site protonation state in aspartic proteases: implications for drug design. Curr Pharm Des 19(23):4257–4275
33. Gossas T et al (2013) The advantage of biosensor analysis over enzyme inhibition studies for slow dissociating inhibitors – characterization of hydroxamate-based matrix metalloproteinase-12 inhibitors. Med Chem Commun 4(2):432–442

34. Poliakov A et al (2007) Mechanistic studies of electrophilic protease inhibitors of full length hepatic C virus (HCV) NS3. J Enzyme Inhib Med Chem 22(2):191–199

35. Doyle JS et al (2013) Current and emerging antiviral treatments for hepatitis C infection. Br J Clin Pharmacol 75(4):931–943

36. Liang TJ, Ghany MG (2013) Current and future therapies for hepatitis C virus infection. New Engl J Med 368(20):1907–1917

37. Geitmann M, Dahl G, Danielson UH (2011) Mechanistic and kinetic characterization of hepatitis C virus NS3 protein interactions with NS4A and protease inhibitors. J Mol Recognit 24(1):60–70

38. Llinàs-Brunet M et al (2004) Structure – activity study on a novel series of macrocyclic inhibitors of the hepatitis C virus NS3 protease leading to the discovery of BILN 2061. J Med Chem 47(7):1605–1608

39. Jiang Y et al (2013) Discovery of danoprevir (ITMN-191/R7227), a highly selective and potent inhibitor of Hepatitis C Virus (HCV) NS3/4A protease. J Med Chem 57:1753–1769

40. Seeger C et al (2012) Kinetic and mechanistic differences in the interactions between caldendrin and calmodulin with AKAP79 suggest different roles in synaptic function. J Mol Recognit 25(10):495–503

41. Seeger C et al (2012) Histaminergic pharmacology of homo-oligomeric beta3 gamma-aminobutyric acid type A receptors characterized by surface plasmon resonance biosensor technology. Biochem Pharmacol 84(3):341–351